Wine from Sky to Earth

Growing & Appreciating Biodynamic Wine

Wine from Sky to Earth

Growing & Appreciating Biodynamic Wine

by Nicolas Joly
Translated by George C. Andrews

Acres U.S.A.
Austin, Texas

Wine from Sky to Earth

Growing & Appreciating Biodynamic Wine

Acres U.S.A., Publishers
P.O. Box 91299
Austin, Texas 78709 U.S.A.
(512) 892-4400 • fax (512) 892-4448
info@acresusa.com • www.acresusa.com

Originally published in France by Éditions Sang de la Terre, under the title *Le Vin du Ciel à la Terre*.

Photos on front cover and pages 88, 94 and 95 by Guy Fleury, used by permission.

Publisher's Cataloging-in-Publication
Joly, Nicholas.
    Wine From Sky to Earth : Growing and
appreciating biodynamic wine / by Nicolas Joly ;
translated by George C. Andrews. — 2nd English ed.
    p. cm.
    Includes bibliographical references and index.
    ISBN: 0-911311-60-2

    1. Viticulture. 2. Wine and winemaking.
3. Organic farming. 4. Agricultural ecology.
I. Title.

SB389.J65 1999   643.8

Library of Congress Catalog card number: 99-073059

Printed and bound in China.

Dedication

This book is dedicated to my parents, who entrusted to me the Coulée de Serrant vineyard, to my children and my wife, to Xavier Florin, principal initiator of biodynamics in France, to Maria Thun, whose work is being understood more and more, and to Rudolf Steiner, whose writings gave a profound meaning to my life.

Contents

Acknowledgements

In particular I wish to thank France Bigourdan for her valuable collaboration in structuring the final version of this book, as well as Editions Triades, the Kepler Institute, and the L.A.P.A.T. (Laboratoire Associatif Pour l'Application des Tests Sensibles) for the documents they so graciously put at our disposition. Special appreciation also goes out to George Andrews for his masterful translation into English and to American biodynamic consultant Alan York for his technical reading of the manuscript.

Preface

Biodynamics is often considered to be a marginal, even eccentric, way of doing things. However, biodynamics is almost entirely based on ancestral methods. For example, we acknowledge the influence of the moon, sun and stars on the tides, but at present we attempt to deny that the same effects extend to grapevines, which we treat like machines that are turned on or off according to our wishes. There are many reasons for which biodynamics could be applauded, but central among them is its abandonment of all chemical fertilizers, pesticides and fungicides. These substances may have disastrous side-effects and this much is certain: they change the nature of the soil and make nonsense of the concept of labels of origin.

It seems to me that a grape in harmony with its environment produces a wine with a more complex aroma, which after all, at a time when wines resemble each other more and more, should be of great interest to lovers of good wine.

— *David Ridgway*
1st Sommelier
La Tour d'Argent
November 29, 1996

La Tour d'Argent is a Michelin three-star rated restaurant in Paris.

"If you want to approach the infinite, approach the definite, but from all sides." — *Goethe*

"One of these two ministers knew everything and understood nothing. The other knew nothing and understood everything." — *de Gaulle*

"He who is only a soldier
is a bad soldier,
he who is only a professor
is a bad professor,
he who is only a manufacturer
is a bad manufacturer." — *Lyautey*

Introduction
Rebirth of the Coulée de Serrant

My childhood and adolescence were enhanced by nature. Born with a passion for fishing and hunting, my brother and I spent our time tracking game, watching for fish, and watching the sky. Each burst of wind, each drop in temperature permitted us, according to the seasons, to make a choice of future expeditions. After a few years of study in the United States and the beginning of a career in the financial sector, I decided to break with a future that was considered promising and to establish myself on the family vineyard, beside my mother. The property was maintained according to the old methods. Chemical fertilizers were used only rarely and in small amounts. Insecticides had never been used on the Coulée de Serrant. Confident in my "progressive" education, I greeted the first visit of a government agricultural consultant with a wide-open mind. He pointed out to me rather bluntly that the management of our vineyard was, to say the least, archaic. We should proceed in a rational manner, by introducing herbicides, which would on their own bring about substantial savings, support the harvest by different interventions, etc. Modern methods entered by the front door. Hard work came to an end. The horses which worked in a nice arrow-head formation to overcome the steepness of certain slopes remained lazily in pasture. After two years of chemical agriculture I gradually became aware of a change in the color of the soil; it was compressed and erosion had caused real gaps at the roots of the vines after heavy rains. The animals and insects had disappeared. I felt that I had installed a vacuum, and was assisting in the destruction of a harmonious whole.

It was completely by chance that I bought my first book on bio-dynamics which I found on the shelf of a secondhand book store. The cover had attracted my attention. The book fascinated me from its first pages. I read and re-read it for several weeks before launching myself on my own into the work of reconversion. On my own, because I was at the time convinced that the ideas of this unknown author had never been put into practice. Thus in France, the biodynamic farmers or winegrowers I met a year later were very few in number, in contrast to Germany, which has always been ahead in this respect. Afraid of taking too large a risk, I at first limited my experiment to a small area — a third of a hectare (less than one acre) — which had been planted in Cabernet vines, much less prestigious than the Coulée de Serrant. The results rapidly convinced me. Three years later, in 1984, the 12 hectares of the vineyard were converted to biodynamics.

Little by little, I saw nature being reborn. The soil was breathing. Autumn and spring were once again periods of hard work. Ladybugs and other insects returned among the vines. But this time, each of us was conscious of their presence. The estate had escaped from a very great danger, a deceptive evil, concealed behind the veil of false progress. From season to season, our understanding of the activity dictated by biodynamics was fine-tuned. Of course, there is still a lot to learn. But from vintage to vintage, the results obtained have confirmed this intimate conviction. We are on the road of truth.

It was in 1924 that Rudolf Steiner, while the guest of Count Keyserling, prepared the foundation for biodynamic agriculture in his lectures, now compiled as the book *Agriculture*. A pioneer branch of biological agriculture, biodynamics was officially recognized by the French government in 1987. Its Demeter trademark is protected worldwide. However, biodynamics was only applied to winemaking after a long delay, even though along with per-fumes, wine is one of the areas of agriculture in which quality — a major focus of biodynamics — is the most studied, even diag-nosed, by professional tasters and wine lovers. Furthermore, there is no book in existence which permits the discovery of this approach to winemaking for those who find Steiner's work difficult to understand.

That is why, after fifteen years of practical experience and at the request of people participating in the conferences I have given, I decided to transmit as simply as possible this experiment and what has been learned from it in a work that tries to bridge the gap between the quantitative visible world accessible to our physical senses and a qualitative, more subtle world which is the source of life.

This book does not pretend to teach anything to those who have already practiced biodynamics for many years. It is intended for winegrowers who devote themselves to their vines with passion and who are now beginning to understand that the advice (sometimes highly priced) which they have been given for several decades constitutes a danger and goes against their interests and objectives. This book is also intended for all winegrowers who want to understand the winegrowing methods of the new millennium, after having seen through the mirages of false progress. And finally, this is a book for wine lovers, so they may better understand the grape.

Historical Background

The Coulée de Serrant vineyard, created in the 12th century by Cistercian monks, has been growing vines for over 800 years. During the same period, a fortress was built a few hundred yards from it on a rocky spur that overlooks the Loire river. It is also here that the famous battle took place, in which Philip Augustus, the King of France, fought the Plantagenets and saved the integrity of the kingdom, taking back certain territories which had been conquered by the English. The property retains the imprint of this period of history. The house, which was rebuilt at the end of the 18th century, extends to the ruins of the fortress. The great alley of cypress trees which overlook the vineyard and the Loire bears the name of the English cemetery, in memory of this ancient battle. As for the monastery, it is still standing and contains part of the cellar. The wine is aged within these walls of historic significance. The Coulée de Serrant, a quality control label of origin that covers an area of seven hectares, has been considered a privileged terrain for over 800 years. It was visited by King Louis XI, Louis XIV, and the Empress Josephine. This wine, which is famous worldwide, was considered by Curnonsky, the prince of gourmets, as one of the five best white wines produced in France.

Chapter I
False Progress

"Matter is nothing, what counts is the gesture
that made it." — *Goethe*

Biodynamics was applied to wine growing long after it was
used in other areas of agriculture. This is, after all, surprising since,
out of all the products of agriculture, wine and perfume remain
the two areas in which quality — a major goal of biodynamics —
receives the most attention. One smells, tastes and chews wine
without ever becoming tired of it. It is thus the responsibility of
winegrowers — the true ones, the great ones, the truly great ones
— to put in place an alternative to the conventional doctrine pro-
pounded by winegrowing consultants as the only way to success.
It is urgent to put an end to the dishonest conduct which has
enslaved farmers and some winegrowers to the agrichemical
industry and the obligatory gifts of taxpayers, whether they be
direct or indirect.

An agricultural approach that can be called artistic is the only
genuine way to make a label of origin really significant. But have
we forgotten the original meaning of the quality control label of
origin?

Created in the 1930s, the first labels of origin were the result of
an intimate knowledge of the varied terrains in the wine regions
and were founded on the observations and experience of several
generations of winegrowers. This experience had brought about

the union of specific types of vines with particular types of terrain. From such legitimate marriages were born wines whose expression was original because they were intimately linked to their environment and therefore inimitable. In this context, the Appellation d'Origine Contrôlée (quality control label of origin) had no other objective than to protect and recognize this originality by offering it a legal existence. At that time agriculture was still healthy and no one imagined a necessity for legislation to guarantee its quality.

Sixty years later, the situation has evolved dramatically. The right to a label of origin has been reduced to its simplest expression: an identification number, a few feeble restrictions on quantities harvested, and a list of authorized types of vines — restrictions which, no matter how light they may be, are nevertheless circumvented. Whether the soil is living or dead, whether the earth is restored or not — fertilized by compost or by the garbage of Paris — whether the taste comes from the earth or from one of the innumerable yeasts that our industry knows so well how to produce, the right to the label of origin remains the same. The routine of wine tasting has a charm of its own — a subject it is preferable not to dwell upon — that is supposed to oversee a selection. In a few decades, the respectability of an inspired creation has been almost entirely destroyed. Wine, now a child of technology even more than a child of nature, flatters the palates of the neophytes, and thus the greatest number. The consequence of this dissolution of the idea and even the *purpose* of the label of origin did not take long to become obvious. Our vines have lost their virtue of originality that is linked to the special quality of a place. They are easily imitated by all the industrialized nations of the world capable of employing the same techniques, while paying workers lower wages.

A few small groups of genuine winegrowers survived the qualitative drama of the labels of origin. Among them it is not only still possible to make wine as in the old days — sometimes grandiose, always sincere — but also and above all to approach what could be called "the wine of tomorrow." To attain this true quality in the product, it is necessary for us to help the vine intensify the manifestation of the earth, of the atmospheric and solar world which composes its environment, and to improve its photosynthesis, which cannot be reduced to automation.

Flight in Advance

This qualitative drama was started off by the introduction of herbicides, a so-called miracle product which made it no longer necessary to work the earth. Herbicides appeared among the vines during the early 1960s. At first winegrowers greeted this novelty with an attitude of prudence, using only small amounts on a few sample areas. This vigilance, not to say distrust, is without doubt an indication of the lucidity of the old-timers who intuitively perceived the falsehood concealed in the speeches about the benefits of herbicides. Promising results overcame final resistances. The technical consultants and other salesmen in white coats, who are paid off in sales commissions, use an argument that packs a punch — lower labor costs. This evil spread all the more rapidly because, after the first treatments, the vines seemed to be doing well. Sample areas were looking better than surrounding areas. However, this phenomenon is easily explained. As the herbicide was applied to the area, the living organisms in the soil died, providing a temporary compost. But simultaneous with this temporary boost, the herbicide slowly but surely weakened the earth. Reserves of health and vitality — built up for decades, sometimes centuries, in agriculture — can be emptied like the balance of a bank account is emptied when no deposits are made. It is really our "soil account" that is being squandered. This "soil account," which has freely encouraged the growth of the vine since time began, holds the key to the specific qualities of our wines.

In this respect one must understand how a soil "works" — it is essential. No roots can feed any type of plant without the assistance of soil microorganisms. Each type helps the root to assimilate one aspect of the geology of the soil. By way of comparison, on a well-set table, you may find a fork shaped for fish, another for gherkin or snail or oysters, etc. Each shape is adapted for the food we want to eat. Each species of microorganism acts a little bit in the same way, permitting the roots of a vine to assimilate a different aspect of the soil. In their absence, the root will starve, being incapable of feeding itself on its own.

The use of the first herbicides marked not only the
beginning of an impoverishment, but also the first step
toward dependence, increased
each year to the point of
enslavement, on the agrichemical industry.

After five or 10 years of herbicide treatments, the free
and generous growth offered
by nature becomes more and
more diminished. Faced with
the anxiety of the winegrower, the consultant opposed it
with a reassuring smile. "Grow
your vines, don't worry, we
know how to do it. In our
days, the soil is of no importance. It is a lifeless support
medium, nothing more."

*A vine which began receiving weed-
killer at 7 years old exhibits roots that
have grown to the surface after 8 years
of weed-killers.*

That, unfortunately, is not an exaggeration. The winegrower
opens the way little by little to a chemical market that has turned
out to be highly lucrative for the agrichemical industry. What's it
all about? Quite simply, the goal is to replace the natural growth
of the soil with chemical fertilizers. These salt fertilizers have completed the work of destruction started by herbicides and suppress
the wee bit of life remaining in the soil while dragging the winegrower even further off balance. These chemical fertilizers force
the plant to gorge itself with water to compensate for the salt. In
soil that has died, the roots of the vines no longer find any resonance, any response to this qualitative solar world which is active
and prolonged beneath the earth through the intermediary of living organisms. The soil becomes deaf to the world which brought
it to life. The first steps toward desertification take place. The
winegrower is no longer worried about nourishing his soil; he
worries only about his vines. Now there are even vines that grow
outside of the soil. And our proud consultants, strong in their

brand new science, tried to convince us that the soil had nothing to do with the label of origin.

The roots of the vines in this new atmosphere come back up to the surface while awaiting their fix (intravenous injection), which later could take the form of foliar fertilizers. The vine no longer expresses itself in its verticality from below. Handicapped by this assistance, roots lose their ability to feed themselves and generate microflora because the microorganisms that would have helped them in this task no longer exist. The imprint of the terrain is progressively replaced by the technology of the cellar.

If you ask a specialist in fauna or subterranean microfauna, he will without hesitation confirm for you the existence of important differences in the composition of the soil from one continent to another, differences still very significant from one country to another, and others more subtle from one region to another, even from one parcel of land to another parcel. Everywhere the earth has a different face. Recognition of the specific nature, if not the personality of each parcel of land, of each type of soil, is at the conception of the birth of the notion of a quality control label of origin. This wealth is being wiped out by so-called progressive agriculture with the benediction and sometimes the encouragement of political leaders.

But the worst is yet to come, for the trap has been sprung on us. If farmers all decided to stop the chemicals now, the planet would go into a state of famine. From the point of view of marketing, it is a master stroke. The chemical industry has put agriculture into a position of total dependence. This would be the lesser of two evils, if we were not aware of the bad effects of this delayed-action bomb. The list of secondary effects of herbicidal and pesticidal treatments available today — though deliberately suppressed — cannot leave anyone indifferent: residues in food, loss of quality, insurmountable pollution (especially of water), and the new plant, animal and even human diseases which result. From now on the obvious necessity for a complete turnabout is directly opposed by the laws of the economy which have become tyrannical. And is it not the absolute limit that to get out of this trap, industry proposes to us that we leap with our eyes closed into genetic manipulations?

The nadir of imbalanced wine growing was reached with the practice of disinfecting soils. All the mistakes made generated nematodes in the soil, which caused great disturbances. To destroy nematodes, the shock treatments prescribed are completely effective. Proof? The destruction of everything alive in the soil is guaranteed after application. But on soils thus disinfected — dead soils — pathogenic elements are the first to reappear. This is what agronomist Claude Bourguignon, who after 10 years at INRA (Institut National de Recherche Agricole) created his own soil analysis laboratory, explains. The consumer has a right to ask what remains of the concept of the label of origin — aside from the label that adorns the bottle — on land so mistreated?

The Soil is a Living and Receptive Organism

To understand a soil and retrieve its character, one must feel attached to it as a *living and receptive* organism, as Mother Earth, as the old ones said correctly. Soil should be inhabited by life. It can contain up to a billion living organisms per gram, as Claude Bourguignon repeats to us over and over again in his book, *Le Sol, La Terre et les Champs*. And that is not all. This swarming life of infinitely small beings is not mixed at random, but, on the contrary, admirably organized in a chain of life where each link allows another to exist. One goes thus from the superficial level of mushrooms, to the more profound one of microbes and bacteria. Another essential point is that these living organisms are always different, not only according to the geology of the soil, but also, and maybe even above all, according to what happens above the soil. That is to say, the climate or microclimate, the orientation of the slopes, the landscape, dominant winds, etc., each one of these multiple living elements acts directly or indirectly on our vines and on the originality of their production. To live in this zone of interface between the solid and atmospheric worlds is an essential step toward understanding biodynamics better. In our time, scientists concentrate their research on the elements taken separately, without seeing the "gestures" which created them, the movements of exchange from which they came and which surround them. Examples of the limitations of such a scientific approach are numerous. Is it not troubling to realize, for example, that the vitamin C contained in wild rose hips is much more active and effective than the little pills of the same vitamin that have been synthesized?

The elements cannot be separated, but on the contrary are linked with each other. Their synergy permits them to express themselves completely. Thus we find again the etymological sense of the word "religious," which derives from "relier," which means "to link together." Goethe wrote on this topic, "Matter is nothing, what counts is the gesture that made it" (*Metamorphosis of Plants*). In other words, for the study of the soil, it is not only the living elements which should be taken into consideration, but also and above all the surrounding world which created them and favored their appearance. Thus the same matter acts differently according to each of the impulses it comes from. Lithotame, dolomite, clay, oyster shell and bone show us different forms of lime.

At this stage, it is important to warn winegrowers who, after exposure to intelligent advertising, buy bacteria to vitalize their soil. These bacteria die rapidly and are not able to make a real difference. Only the living surroundings which inhabit the soil are capable of generating them and, when it is absent, they are incapable of reproducing themselves. It is not the bacteria which make the soil, but the quality of the soil which permits the appearance and reproduction of bacteria.

In regaining consciousness of the interference between earth and atmosphere, we learn to see what descends into the soil and what animates it. If the Earth were separated from its solar system by an immense sheet of black plastic — the saturation by electromagnetic waves may well constitute this opaque screen in another form — life in its major forms would rapidly disappear. It is thus essential to carefully improve the receptive capacity of our soil and to orient this receptive act in one direction or another, for example, by the choice of different manures. But before discussing this agricultural activity and to understand thoroughly and in depth these interactions between sky and earth, we must first of all examine the different states of matter.

Chapter II
The Great Balance: The Four States of Nature

The earth is much more than a solid body beneath our feet; it extends hundreds of miles above our heads. This less visible part of the earth is the scene of a series of exchanges that form a true link between that which is part of the earth and those things that do not obey terrestrial laws. Here are woven connections between the laws of life and those of matter. This area, full of life in all senses, is where a subtle reality from a faraway world is melded into the reality of our own life system. This delicate system of exchanges places a great responsibility upon mankind, for if such exchanges are no longer possible, the very survival of our world is threatened. Everything must be harmonious here — the *Harmonia Mundi* — as recognized by Johannes Kepler in the 14th century, as well as by ancient philosophers such as Plato, etc., who wrote of the "music of the spheres" (planets). This harmony is not a matter of superstition or fantasy, but of perception. In an age when society did not yet tyrannize our senses in order to destroy them, perhaps man had a finer perception, an ability to respond to such harmonies. In our day, wine tasters have demonstrated their ability to sense odors and flavors. This appears to indicate that we have the ability to develop finer sensibilities in other domains. To send harmful things into our atmosphere — such as waves of energy — is surely not a neutral or inconsequential action. A wave of energy can destroy other forms of energy, a fact recognized by the World Health Organization, which has devoted an entire report to this subject. We must take great care not to turn the harmonious music of the spheres into a dissonance — the risks are too great. The earth is one component of the solar system, and within that system, communication (or a sort of balance) is accomplished through wavelengths or frequencies of many sorts. Life itself is only made up of frequencies or microfrequencies.

Biodynamics is capable of revealing this more complex perception of reality by approaching the life that surrounds us in a new way. Biodynamics, for example, recognizes the four different states of matter. Moving from the most solid to the most subtle, these are mineral, liquid, gas (where light is manifested) and finally heat. This is not a new concept; Aristotle himself recognized these states. And medieval botanists always characterized by one or two of these four qualities, using the terms hot, cold (terrestrial), wet or dry (light). Thus, Hildegard de Bingen in the *Book of Subtleties of Divine Creatures*, described the plant valerian as "hot and wet" and thyme as "hot and dry." In a manuscript of the medieval period housed at the Bibliothèque National in France entitled *Book of Simple Medicine*, sage is considered "hot in the first degree and dry in the second," while mushrooms are "cold in the first degree." Nicholas Culpeper, the most famous of the English botanists in the 16th century, expressed himself in similar terms.

The Mineral State and Roots

What is the mineral state if not the hard part of the earth, the rocks and stones? The earth, understood in the sense of "terroir" (terrain), as well as all that is above and around it, is a more complex idea that leads us to a higher stage. At the level of the first state of matter, the forces of gravity are dominant. These forces make things heavy, hard, condensed, draw all toward the center of the earth, allow matter to appear and cause elements to coalesce. The mineral state is very solid, almost petrified, a place where life is, for the most part, nonexistent. This state is only part of our planet; in a way it is the skeleton of our planet.

A plant is related to gravity and the mineral state mainly through its roots. They plunge toward the center of the earth. Their power is centripetal, turned toward the interior, thus heavy and hard. To understand the mineral state, take in one hand a vine root and in the other a vine shoot. Do not analyze the differences between these two kinds of life, but instead try to feel how the forces in action are different. In our time, the vine needs us to understand these different states of its existence.

The Liquid State and Leaves

Above the mineral state another one exists, a more mobile, subtle state — liquid. Water is, of course, the archetypal example. Unlike the mineral state, the liquid state is not fixed into an unchangeable form. The content of two glasses of water can be mixed; we cannot do that with two stones! When a person enters the water for a swim, she feels lighter, less terrestrial, in the sense that she is a bit delivered from gravitational forces. This is the well-known principle of Archimedes. For that reason our very brains are bathed in a liquid state to allow us to think; an embryo survives in such a state until it is ready to be born. This liquid state allows our food to be assimilated by our bodies.

Even though it, too, is subject to the laws of gravity, water retains a certain independence. It acts as a kind of intermediary between two different worlds. When frozen, it becomes solid, almost mineral, a prisoner of matter, and fixed in a certain form. With heat, on the other hand, water becomes light, airy, and rises in fog and mist. But it never crosses an essential barrier. When it reaches a certain height, it falls again as rain, filled with subtle elements. How different from water as it flows from our faucets.

In plants, the liquid state is manifested by sap, mainly, which acts for the plant as a link to sun forces! Let's make sure we do not disturb that link.

The State of Light and the Flower

Light is an airy state intimately linked to our atmosphere. This atmosphere that surrounds the hard kernel of our earth is itself divided into separate layers: the stratosphere, the troposphere, the mesosphere, the thermosphere. This third state of matter, more subtle and less visible than the first two, is manifested in light. Airy, gaseous and luminous, the state of light surrounds us continually and circles our globe, almost completely free of the laws of gravity.

Concerning plants, the fragile state of light is reflected in the flower with its color and finesse. Each plant possesses its own level of ability to catch or to link itself to this third state of matter. Few reach the level of the lily, where light forces have been caught so intensely. Others, we will see, direct these light forces away from their flowers and to their roots — for example, carrots. Unlike roots, the flower's energies are centrifugal, turned toward the exterior. See how refined this state of matter is. The flower shows us a real dissolution of matter in its odors and almost imperceptible pollens. It is these delicate qualities that the wine-makers hope to concentrate in their wines. To understand the scope of the phenomenon, Goethe's *Metamorphosis of Plants* is an essential guide. Goethe shows us the secret mutations within a plant, the exchanges between its roots and its flowers.

The State of Heat and Fruit

The state of heat is the least terrestrial of all the states of matter. It flees from matter, or tends to make it dissolve or disappear. It dilates matter. A hard piece of iron exposed to enough heat becomes liquid; when exposed to even more heat, it becomes a gas. Matter has "disappeared." Life requires a minimum level of heat — each form of life on earth is a balance between these four levels of matter, and heat can also generate life. In the following chapters we will explore how it is possible through biodynamics to reinforce the action of heat (in northern regions) or to slow it down (southern regions).

"Light and heat are the indispensable conditions for plant life. . . . Light is the essential material from which come agricultural products, and heat is the force that activates the mechanism of the plant," wrote W.R. Williams in 1952 in his *Principles of Agriculture*. The energy of this state of heat, when expressed within reasonable limits, allows for the fertilization of the flower. Thus once again a flower takes on matter in a new way and eventually evolves into a seed and a fruit. During every day of this process we can see forces of gravity act on the fruit as it grows, until the fruit with its seeds falls to the ground. After a period of hibernation during the winter, the seed can germinate and come forth with new life, to rise out of the earth. This hibernation state is interesting because it reveals how the seed is acted on by subtle principles that need a certain material support in order to function. The seed thus marks, year by year, a period of renaissance. It can bring forth new grape vines that, if they are well created, will be regenerated, full of fresh energies, and more resistant to pests and diseases such as phylloxera.

The state of heat is then the force that gives birth to fruit. Independent of its regenerative action, it is a state of energy that rises. A hot-air balloon is a perfect example of its action. This fourth state of matter is concentrated in the thermosphere, the hottest and farthest from the earth of all the levels of our atmosphere, forming a veritable belt of heat around us, like a womb. Hydrogen is closely linked to this state.

The Games of the World of Vegetation

It is essential to recognize the four states of matter in the plant world, in order to understand the dual movement that is taking place within each plant. One movement is toward the bottom (marked by the laws of gravity), thus toward more intense matter (roots) and another movement is going up at the opposite extremity of the plant, with a beginning of dematerialization (odors, pollen) and a tendency toward lightness that favors the rise of branches.

Gravity and levity are a polarity of opposing forces that have many expressions in the world of vegetation. Each tree and each plant integrates these forces in a distinct, almost personal fashion. To understand this, one has only to look at what surrounds us. Each plant has a special significance. Plants communicate distinct

The cedar tree is very well balanced between upward and downward forces, with branches that display almost perfect horizontality.

A willow is strongly linked to water, with branches that fall after an attempt to rise.

Vines are strongly linked to earth forces and thus to gravitation.

Cypress, strongly linked to heat forces, climbs upward.

messages if only one is open to hearing them: one plant will rise toward the sky, catching more of these ascending forces; another, receiving more gravitational forces, will plunge toward the earth, piercing a harsh soil with its roots; and a third will maintain a balance between these rising and descending forces and show a perfect horizontality.

First, consider the earth-sun axis and compare it to the form of a cypress, which is characterized by almost total verticality. Some varieties are especially notable for being little more than narrow tubes rising straight to the sky, with no horizontal growth. One can see how much these trees are linked to the sun and why they are planted in cemeteries. These types of plant were called *Apollinian* by the ancient Greeks. Apollo was the sun god, so these were "plants devoted to the sun." Wheat and laurel are just two among many Apollinian plants. The cedar, on the other hand, expresses both movements in almost perfect balance. Its movement is completely different from that of the cypress, with branches that seem almost to be suspended like the wing of a bird gliding through the air. Here the forces of gravity and levity are in perfect equilibrium. The weeping willow, in contrast to the two previous trees, seems almost weighted down. Its branches quickly lose their ability to rise, and fall back to the earth filled with water (this tree can absorb up to 4,000 liters [1,140 gallons] per day). The tree is completely unlike the rigidly vertical cypress.

Although the four states of matter are found in every plant — mineral predominating in roots, water in leaves, light in flowers, and heat in fruit — they do not always manifest themselves as we might expect. Nature, luckily, remains free.

Some trees concentrate forces of light in their wood to create, for example, the bright yellow of wicker or the white bark of the birch. Leaves are the same. In autumn when the sap falls, the maple — symbol of Canada — reveals colors that cannot be seen in summer, but in the spring we can taste a new state of the tree in its sugary sap. This underlines its special capacity for attracting light and heat forces, acting not on its flower, but on its leaves and sap. The silvery back of a poplar leaf also manifests the action of light.

Heat often works with light to create, at different stages, a particular effect. Thus an odor, more commonly associated with the flower, permeates even the trunk of cedar or sandalwood trees. The hardness of the mineral state can make some woods, such as that of the service tree, so hard that they discourage the woodsman's ax and the carpenter's tools. The pine has floral qualities in its wood and mineral hardness rises to the level of the flower to create the pine cone.

These complex actions in the plant kingdom were once recognized by man. Understanding these factors again is indispensable for anyone who wishes to cultivate the vine and truly comprehend how the plant kingdom works. In each plant there is a sort of secret momentum which can be used for supporting the quality of other plants. It is indeed true for vines. Two books are very illuminating in this respect: *Man and Medical Plants*, by Wilhelm Pelikan, and *The Plant: An Approach to its True Nature*, by Gerbert Grohmann. These two polarities — matter and nonmatter — or the four levels of matter can and have been extended into many fields.

Our four "tempers" (the ancient concept of the humors) each show one of these four levels dominant over the others. Choleric (heat); sanguine (light/air), which means movement if not agitation; melancholic (earth or coldness); and phlegmatic (water). When you ask a phlegmatic in France how he is, he might answer, *"ça flotte,"* which literally means "it is floating," and figuratively means, "I feel well." Floating means being a bit less taken by the earthly level, so it is a good feeling.

The four levels of matter can be applied to gas, too. Hydrogen (99.5 percent at 150 kilometers of altitude) is behind the process of pollination and the levitation of water (fog). Nitrogen linked to air brings movement, exchanges; oxygen linked to water (much more than in the air) is the carrier of life on earth, or life at an earthly level. Carbon is the physical level. It can combine with itself, and that brings a large diversity of forms for the earth. But carbon can also be a diamond, showing its origin.

All this is always a trip from macrocosm to microcosm and the reverse. A permanent birth-and-death process. One could look at acids and bases in the same way. An acid dissolves everything — it disincarnates, if you prefer. A base is the reverse: it precipitates a substance (incarnation). All this knowledge is available in the works of Goethe, Steiner, Hauschka, Walter Cloos, etc. It is also taught to children in Waldorf Schools.

The Leaf and the Flower — a Metamorphosis of the Same State

Goethe, in his *Metamorphosis of Plants*, explores the continuity that exists between the leaf and the flower. These two plant forms are not two different states, but rather the same state metamorphosed. A descending force meets a force rising from the ground. From their respective powers, a particular quality is created in the flower. Of course, each variety of plant will operate in

The effects of the four states of matter on the plant

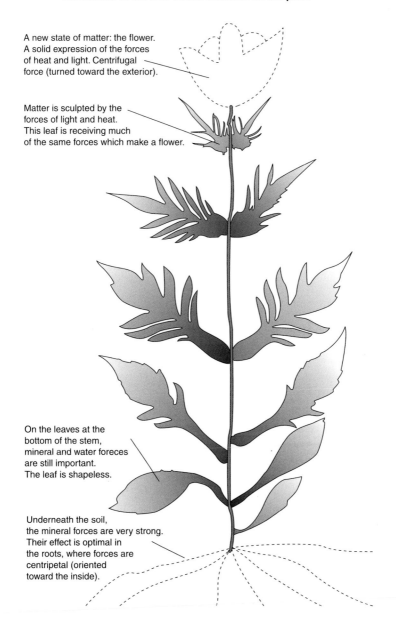

A new state of matter: the flower.
A solid expression of the forces
of heat and light. Centrifugal
force (turned toward the exterior).

Matter is sculpted by the
forces of light and heat.
This leaf is receiving much
of the same forces which make a flower.

On the leaves at the
bottom of the stem,
mineral and water foreces
are still important.
The leaf is shapeless.

Underneath the soil,
the mineral forces are very strong.
Their effect is optimal in
the roots, where forces are
centripetal (oriented
toward the inside).

*Drawing extracted from the book The Plant: An Approach to its True Nature,
by G. Grohmann. Captions by Nicolas Joly.*

its own way, more or less differently from the primary model described by Goethe. Grohmann describes this metamorphosis very clearly. For example, the leaves of the field scabious (*Knautia arvensis*) become more and more ragged, thinner and smaller as they rise. One can say that the mold sculpting their matter becomes more precise as matter becomes less dense. "Counter-forces," cosmic or solar, transform in their descent that which is rising from the ground. Through this double movement a kind of mutation occurs whereby the leaf finally becomes a bouquet of flowers. The forces of light and color predominate over mineral and aquatic forces, favoring the appearance of finesse, sensitivity and fragility in the flower — the flower whose life span will be much shorter than that of the leaf. A flower always captivates us because it is the incarnation of light, this other state of matter, more refined and less terrestrial than the others.

In some flowers color is almost nonexistent, but in others, light is expressed with incredible intensity. The leaves of the lily, for example, become smaller and smaller as they appear on the stalk, calling our attention to a new element — the flower — a flower that is totally dominated by the power of light. This example of two forces almost totally divided in the lily is part of the reason for its great impact on our senses and perhaps a reason it was chosen as the emblem of the kings of France.

This interaction between opposing forces exists in all plants, but we seem no longer able to understand this "language." But the vine grower, by seeing plants in a new way, can reach an understanding of the role that landscape plays in the life of any plant and particularly in grapes classified as part of a particular label of origin — it is this land that in many cases must be brought back to life. As we will see, treating vines with teas made from specific plants according to the season, climate, etc., will support specific qualities of the grape.

A New Way of Looking at Our Vines

The vine plunges into the soil with great force. Almost nothing can resist its roots, which can descend through the tiniest crack in the stoniest soil. They can descend hundreds of feet beneath the surface of the soil. This aspect shows how closely linked the vine is to the forces of gravity — what the Greeks called an "underground force." This is the meaning behind *Dionysus,* as we shall see. In contrast, the vine is not able to rise well and fully use the forces of levity given by the sun. We are witnesses to a kind of drama. The vine is a prisoner of the earth. Without the help of some sort of support, the vine can only dream of rising. In the spring of each year, the vine repeats its difficult attempt to escape into the air. Each time its lack of structure dominates its rising forces, the vine attaches itself to whatever it can to recreate a movement that is condemned to fail. Its powerful descent is not counterbalanced by an ability to rise. A balance between these two forces is achieved by the yellow gentian (*Geniana ornata*), for example, only after seven years of preparation. This plant spends six years preparing a five-foot-tall stalk and a powerful, deep root. Only in the seventh year does the flower appear.

The grapevine, in spite of all its efforts, will never succeed in its quest. Like any creeping plant, it is dominated by a spiral action, turning around an axis without ever attaining its goal. Its role is to bring man back down to earth. In spite of its cosmic origins, the vine is a prisoner of the earth.

Nature nevertheless gives the vine satisfaction in its solar quest. Most fruit-bearing plants begin their vegetative cycle with flowers, which eventually will be followed by leaves. The vine, in contrast, can hold back its flowers until June, the time of the summer solstice, when the days are the longest of the year. This underlines the very special relationship of the vine and the sun. When the vine flowers, it finally achieves what it has sought for so long. It becomes fertile during the time of a summer solstice sun, not a springtime sun, but summer sunlight that is more intense and lasts longer. The vine flower, waiting for summer to appear, is minuscule, tiny, almost hidden within the plant. In contrast to the lily, which is totally turned to the exterior, the grape flower is internalized. Its action invites sunlight to penetrate within the grape stalk. And as if to share its internalization with us, it emits a powerful odor, capable of attracting our attention from a considerable distance. The vine, which has required us to support its outflow, now seems to turn inward on itself after this summer marriage, with a force that is as intense as it is difficult to see. We find this intensity again when the grape juice starts to ferment.

The vine's growth rhythms emphasize its affinity with the sun. The vine follows the sun during its course. From spring to summer, the sun rising in the sky draws the vine to an external world in which it precipitates its vigor. After the summer solstice, however, the sun's diminishing presence makes the vine turn in on itself. The sun halts the vine's growth to focus the plant's energies almost solely on the ripening fruit. The vine becomes wiser. It no longer expresses itself in movement with the desire to conquer space. It is concentrated on its fruit, its seed. Its vegetative cycle is the perfect reflection of its solar origin. Finally, in autumn, the vine offers us its fruits which, if the summer fertilization has been carried out correctly, are heavy, substantial and full of flavor.

From the manner in which a vine grows in a vineyard, one can judge the quality of the agricultural techniques being used. The vine naturally follows the rhythm of the seasons. The winegrower can help the plant through appropriate actions. Thus, at the moment when the vine begins its "inward" journey after flowering, focusing itself on its fruit and seeds, do not obstruct this natural action by artificially forcing the vine to turn "outward" in its growth through the application of harsh nitrate fertilizers that are blind and deaf to the rhythm of the seasons. The shock to the plant of obstructing its natural forces by unnatural ones can cause rot. And to ask wine research centers to create new varieties of grapes that are rot-resistant can lead only to additional imbalance.

Guided by his own observation and understanding of the life of the vine, the winegrower can determine what kind of pruning is necessary and to what height the vine should rise. The winegrower can try to satisfy the plant and its solar affinity by imitating what is done in Portugal, providing a support several meters tall for the vine to climb upon. But in trying to satisfy it, he might put it to sleep or, contrary to his intention he could stunt the vine and send it back to the soil by contradicting it to provoke its prodigiously deep and vigorous reaction. In this domain, as in so many others, there is no universal solution. Our choice is conditioned by diverse and varied parameters, beginning with our own sensitivity to the vine, the qualitative and quantitative objectives, and of course the geographic location and the climate.

As Grohmann points out, the forces of light are concentrated on the two poles of the earth. No one can travel to Nordic countries without being aware of the special qualities of light there. What a surprise to see, in Labrador for example, a carpet of tiny flowers on the surface of the soil. When we examine such flowers close-ly, we see that often they are varieties that exist in more southern locations, but here in the north, the light is so powerful that the flowers have no need to rise toward it, and the force of the earth is so diminished that stalks and leaves remain tiny.

By way of contrast, at the equator we see a totally different growth pattern. Terrestrial forces are so powerful here that they work even above the soil. We see plants attaching themselves to nearby trees and growing with their roots above the soil in the air nourished by terrestrial forces that are normally active only under-neath the earth. Everything grows with gigantism, to excess, but with a lack of refinement and restraint. We are surrounded by a proliferation of vegetative forces and overpowering odors, which would not really be appropriate for wines. Only a certain balance between these extremes can produce a good wine.

Every good winegrower should understand these two polarities in order to situate his own appellation and his own vines. Knowledge of the four states of matter in his particular location will help the winegrower understand how to proceed: must he accompany, accentuate or defeat certain tendencies in the vine? Biodynamics can answer such questions because it does not focus directly on physical manifestations, but rather on that which caus-es these manifestations to occur.

In the Champagne region of France, the vine grows at its north-ern limits. The winegrower in Champagne must work to reinforce light and heat for the vines. In the Midi region, on the other hand, the grower must work to intensify the force of the soil, threatened by drought and an excess of heat.

In Praise of Diversity

We can now understand the particularities of a certain vineyard and see the different qualities the vine can have within one appel-lation, depending on the region or specific location. An appella-tion (territory covered by a label of origin) is a general identity within which individual landscapes are an important element.

The proximity of a vineyard to a forest, a river, a change in altitude, or a variation in wind patterns following the topography of the soil, are all specific factors which each vine expresses in its own way. Each site should manifest its own special qualities. To reinforce the special life of a place is to accentuate diversity and diminish uniformity. But our system of education for winegrowers is designed to encourage uniformity, as well as to privilege quantity over quality. Yet especially for the vine, special and unique qualities should be enhanced, not destroyed for abstract, theoretical reasons. Each winegrower contains within himself or herself a "difference" which can be transmitted to the vines through his or her actions, presence and consciousness.

Biodynamics is a collection of principles which can be adapted to individual circumstances. That is why it is so difficult to speak of biodynamics in general. The understanding of its agricultural bases, its actions, the adaptation of the vine to its particular terrain, are all factors in the quality of a wine. But let us explore this notion of "quality." A biodynamic wine is not necessarily "good," but it is always *authentic*. Biodynamics, by reinforcing the natural life of the vine, accentuates each vine's characteristics. It never favors, for example, the indiscriminate planting of varieties of grapes with no consideration of the location of the vineyard. Quite the contrary. And one must understand this aspect of biodynamics before unjustly accusing the approach of any deficiencies.

Biodynamics, if properly practiced, offers results that are truly remarkable. Biodynamics can become a real saving grace for a vineyard. With only three years of biodynamic cultivation, we have seen vineyards that had been savagely dosed with weed-killers for years begin to show noticeable improvement in health, and quality — improvement that is perceptible in blindfolded tastings of the wines from these vineyards.

To achieve such improvement, vineyard soil must be able to manifest a maximum of life forces. But this life must be specific. How can we maintain and support the originality of a particular vineyard? How can we restore its natural qualities if these have been destroyed? Concretely, what can we do, how and when? For deciding actions, we must now move on to another understanding of the seasons.

Chapter III
Autumn, the Season of Decay

After the autumn equinox the days become shorter than the nights. Obscurity prevails, the descent begun by the sun on the first day of summer takes an important step. Nature feels it intensely. An intense symbiosis reigns between the world of vegetation and the solar system. The results of Professor Faussurier's experiments, presented at a conference at the Université Libre in Lyon, have opened up new perspectives. Hundreds of morpho-chromatographs have allowed him to "photograph" the reactions of many different types of vegetation several hours before the start of a solar eclipse. The image is transformed and returns to its previous appearance 24 hours after the end of the eclipse. Vegetation is part of a link between the sun and earth. Each plant lives out this union at the deepest level of its vital juices; it feels all the solar movements and vibrations. When days are shorter than nights, it means that the laws of the earth are dominating the laws of the sun. Thus, gravitational forces are stronger than solar attraction, or levitating forces, life is slowed down, and most leaves have to fall.

Vegetation is particularly sensitive to the season of autumn. According to the image of the sun, a plant's sap withdraws towards the depths. This is a moment when it is important to work the earth. The surrounding elements prepare to live out a separation — a sleep, a little death. What was brought together in the spring now needs to break up and decay. The leaves turn into humus, into vegetable earth. The way in which things break up is as important as the way in which they come together. In all simplicity, to achieve a compost is to help vegetable or animal matter break down in a way that returns it to a more simple level. The quality of the compost depends on the subtlety and intimacy of these modifications and degradations. Nothing can be neglected — temperature, moisture, light, proportions of vegetable and animal matter, the nature of the animal — everything must be in a correct relationship because this balance permits the retention of certain volatile elements. Once again, it is not only the material which counts, but also its origin and surroundings, thus what it attracts. To force decomposition by artificial methods is something completely different. One wants to retain some of the forces which were active in the plants and animals used in the compost.

To speak of biodynamic compost in terms of alchemy would not be false. All the works dedicated to alchemy, in particular the books of Alexander Von Bernus (*Médecine et Alchimie*) and Thomas Aquinas (*Traité de la Pierre Philosophale, L'art de l'alchimie*), recall the different stages of these modifications of matter, the relation of the elements to each other, their affinities and antipathies, as well as their exchanges during cooking and condensation. To think of the alchemist in terms of making gold is as inappropriate as reducing the profession of growing wine to that of manufacturing rotgut. This profession possesses precious knowledge, founded on the understanding of a distant genesis of elements and substances.

Everything that relates with genuine and sincere understanding to a living background can render great service in restoring balance to increasing disorders which can no longer be cured if one remains only on the material plane.

The Four Realms

To make a compost is thus to preside over a good decomposition. It cannot be a question of bringing together different elements taken at random, making a pile and waiting. Each substance of vegetable or animal origin is consciously selected and assembled according to its tendencies and affinities with (in this case) the vine. But on what criteria do we base our choice? The answers are primarily connected to an understanding of the four realms which surround us: mineral, vegetable, animal and human. Four realms whose complementary nature is obvious.

Inanimate, immobile, silent and sometimes very beautiful, the mineral is not, strictly speaking, inhabited by life. In the realm of vegetation, matter is organized, oriented, alive and more easily assimilated, but it is nonetheless dominated by the external world, having in the final analysis a very weak capacity to escape outside conditions. At the level of the animal realm, an important threshold is passed in the complexity of life. The cell no longer reproduces by simple division but by invagination, by digging into itself, showing its tendency to create an internal space, a center of autonomy, a separate universe. As a figure of speech, we can say that the animal takes into itself a little of the external world in order to free itself from it. This autonomy is the sign of a superior realm. The realm of vegetation may be thirsting to attain this state. Don't certain plants mimic the appearance of animals, sometimes even imitating their movements? These two worlds are intimately linked and winegrowers should be aware of this. Goethe said, "The butterfly is a flower detached from the earth, and the flower is a butterfly attached to the earth."

A man of the earth cannot ignore this exchange between the vegetable and animal realms, their reciprocal dependence in the accomplishment of their daily tasks, their support of each other. The discovery of this unique link enlightens us to the frequently dramatic consequences of the irresponsible use of insecticides, whether they be chemical or biological. The vine is not an exception to this principle. Following the example of other vegetation, it is united to this superior realm, which comforts it. And it derives benefit from the mere presence of an animal.

Above the three realms is man, who stands erect, in vertical rather than horizontal position, an authentic vertical, not to be confused with that of the monkey some people claim we are descended from. No, on the contrary, man's ability to stand erect gives him autonomy, permits him to say "I" with complete liberty, and to have a face that is always different. Only the physical aspect justifies a comparison between man and the monkey "molded" into its race. The hypothesis of a miraculous gene, which would have allowed this mutation, remains unsubstantiated. But this analysis reflects the sickness of our era, during which man, no longer understanding the origin of life, prefers to pay attention to its manifestations. Wisdom would direct us to take the opposite road.

On a winegrowing property, man is the conductor of the orchestra because he is the only one capable of creatively tuning in to the vines by reinforcing sensitivity and by accentuating the exchanges. He should make sure he does not provoke any false notes, that he awakens rather than shuts off.

Observations of the exchanges between these four vital realms is a necessary prelude to a chapter devoted to the nourishment of the soil because only then can one understand why pretending to reinvigorate the soil by adding mineral materials — nearly inanimate matter like lime and potassium — is a procedure which is not very convincing whether or not it takes place under the "bio" label of agriculture. Results are only obtained through the use of massive amounts of materials (several tons) which modify the composition of the soil, thus its life, equilibrium and, of course, the specific characteristics of the appellation.

Feeding one's soil with vegetable matter is already an important step toward life. Communication between two elements belonging to the same realm is always easier to establish. The quantities used can be decreased. Each type of vegetation has its special life energy. It cannot be a question of considering only "vegetable matter" without also taking into consideration, for example, certain affinities. What winegrower has not lived through the marriage of the vine and the peach tree? Those sweet little peaches that quench the thirst of the winegrower. The ease with which a peach pit sprouts when it is surrounded by vines is without a doubt a sign of this reciprocal affinity. This is only one example among many. Others include the mulberry, which works so well as a support for vines. Many such complementarities have been established by scientific tests.

Concerning the animal kingdom, the effects of animal matter on vines are the most pronounced. It is essential for us to regain an understanding of the workings of the animal kingdom in order to correct dangerous imbalances such as the tragedy of "mad cow" disease, to give only one example. This illness was described by Rudolf Steiner in 1923 (see Appendix II). This illness is merely one example of the dangers in store for us if we allow ourselves to be guided by intellect alone.

That which man has caused the animal kingdom to suffer is a debt that will one day need to be paid back. To act on one animal is to act on a whole system of animal life. To change one animal life form is to put the rest at risk. This has been proven in a scientific study in Japan involving rats which were trained to do a specific task. The training took some time, but when the experiment was repeated with another group of rats, they learned more quickly than had the first group, and a third group learned more quickly still. The rats were interconnected, coming from a common "mold," which, when altered, affected all of them. Ignoring this invisible but real background can lead to destruction. Conversely, when we carefully raise an animal, this act can benefit the whole race.

In spite of much evidence concerning the dangers of handling animals improperly, man continues to do so. Certain races of animals have been reduced to mere sources of revenue. Cattle, which can live up to 20 years, are slaughtered at no more than six years. Waiting longer is a risk, because diseases in the animal can make the meat inedible. How many cows would die if antibiotics were not available?

Animal raisers tend to think in terms of how many animals per square meter. To increase the profit from raising pigs, a group of pork farmers advised by the French Ministry of Agriculture had the bright idea of progressively increasing the number of pigs per square meter. The animals began to eat each other's tails and ears, to show clear signs of stress, yet nobody cared and the practice was continued. When the pigs began to eat each other's valuable thighs, however, the farmers stopped the experiment because it had become too costly. The precious hams were losing their value. This shows the priorities of our government.

It is this kind of mentality that governs what we eat today. Can we continue to allow such thinking to dominate the world of agriculture? What person living off the earth could truly believe such practices to be desirable? Can we really be ruled by the thinking

A goat has a strong connection to the upward forces of heat and light; action of the manure is more on the flowering process and perhaps to some extent on taste.

A horse, strongly linked to heat, produces manure that acts on the fruiting process and is also very beneficial to taste.

subject or we will face more disasters like Mad Cow disease.

Thus we must choose the animal or animals whose manure is most appropriate for our particular vineyard. If the land has been in vines for a long time, one can study what types of animals were raised on the property in the past. Talk to the old-timers among the winegrowers, the ones who plowed their land with horse-drawn plows and lived the life of their land in their guts and in their blood. Over the course of time, such farmers developed an acute sensitivity to nature, something which cannot be achieved by intellectual methods alone.

A farmer might tell you that the vines "feel" the presence of a horse, which makes sense when one considers how totally different from the horse-drawn plow is the action of the tractor. To study former practices in this way is not to take a step backwards, but rather to discover the best course of action to follow.

Instead of making use of ancient knowledge about agriculture, and seeing how to improve it, we laugh at these old beliefs. We make fun of such ideas because they are too profound for our mentalities. We have been, in a certain sense, extinguished. Much of the blame can be directed at agricultural training, which teaches us to nourish an animal as a piece of meat without understanding that the animal's diseases can come from within itself when it is treated improperly. This is a destructive approach. How can we understand a subtle world from such a materialistic perspective? When we see a work of art, we cannot appreciate it merely by examining its

The pig is dominated by earth forces — a pig can dig in soil to find roots. Its manure chiefly acts on the growth of roots.

The cow is dominated by liquid forces; its manure will primarily act on leaves.

weight, its thickness and the proportions of its colors. We can be accurate about these things, but we will have completely missed the point.

The Components of Compost

The proper composition of compost must be determined by the specific requirements of each vineyard. The essential factor is to select the proper animal and vegetable matter to be included. One example is cow manure mixed with straw from rye, wheat or barley, to which grape brandies (not distilled) can be added. Since much of the straw available today comes from hybrid or genetically modified plants, the effects of this straw on the four states of matter are difficult to anticipate. It is best to avoid straw sprayed with insecticides or loaded with nitrates which destroy that brilliant gold color of natural wheat so beautifully captured by Van Gogh.

The feeding of animals whose manure we will make use of is also extremely important. Feed filled with antibiotics will dramatically alter the quality of the animal's manure, which will be made almost antiseptic. Animals must be nourished properly, as Rudolf Steiner always insisted. An Austrian veterinarian, Dr. Sellinger, was a world authority on this subject. Who stated that, a bovine's daily intake should consist of one-third leaf matter (cabbage, for example), one-third flowers and fruits (high-quality hay) and one-third roots (such as beets). This illustrates once again the synergy between the four states of matter. The roots work on the head of the animal, the flowers on its metabolic processes, and the leaves on its respiratory and heart functions, this in the middle of two extremes. Certain studies, cited in the appendices to this book, have covered this subject in great detail. Here, I want merely to point out the importance of proper nourishment, especially in our time when qualitative famine is the trademark of our era. Proper nourishment is not only physical substance which is quickly rejected, but forces and energy which draw in matter. This is what quality is about.

The system of nourishment for bovines just described was widely used up until the time of "progress" with its opportunities for artificial feeds. Today the best manure comes from animals raised and fed on the domain. This advice is based on experience, but is sometimes difficult to follow. Thus the INAO (France's regulating board for food and drink) noticed that Reblochon cheese lost its character when the milk cows were fed hay imported from the south. The subtlety of the four states of matter in the mountain hay was not there. The milk was affected and the special taste of the cheese was gone.

These different qualitative interactions are not detectable at first glance. From this point of view, vineyard agriculture cannot be limited to quantitative measurement, to a calculation of averages that tends to melt our product into a mass of uniformity. In other words, multi-crop farming is the ideal way. The interactions that it triggers are a source of infinite richness. And if it is not always compatible with the growing of vines, it is possible to take a step forward by bringing to one's estate at least one domestic animal, to understand and break through the frontiers of monoculture. Biodynamics can provide other answers to the situation of monoculture if it cannot be changed.

The more we approach the agricultural organism that each estate should be composed of, the more we reach conclusions in total opposition to those of the experts. The Ural region was supposed to produce all the wheat for Europe in the years to come. This purely technological prediction, based on a quantitative analysis of agriculture, will turn out to be absurd. Far from an immense monoculture whose lack of balance is temporarily neutralized by applications of toxic substances, we are going to see the return — if we escape from genetic tampering — of small, multi-crop farms and the qualitative, economic and even social balance they enable us to meet. The only way to control disease is a return to harmony.

Preparations for Autumn

We have discussed compost for the soil and the fact that autumn is the most favorable time to put it in the earth. It is now time to talk about the biodynamic approach to compost. These so-called "preparations" give it a whole other reality if they are done right. They are the master key of biodynamics, its very essence and the source of its increased effectiveness. Let's speak again of the winegrower who, in three or four years and in spite of having used herbicides for fifteen years, restored intense life to his soil at a depth of 5 feet. Such a change is incomprehensible if one looks at it only from the physical point of view, so how is it possible?

Soil is thick, heavy and packed down; we think of it as a kind of physical barrier. However, in reality the soil is crisscrossed by an invisible network that is not governed by the physical state of the soil, but can affect it. The life of the soil is born through this network. Each preparation acts as a magnet, to bring together specific life forces.

We will not cover all of these techniques here. That is not the aim of this book. Many specialized studies have covered them. Chief among these is, of course, Rudolf Steiner's *Agriculture*, which is rather difficult to follow. Other works include *Plants for Biodynamic Preparation* by Werner Simonis and *The Biodynamic Preparations* by Koenig, to mention only a few of them.

The plants that Rudolf Steiner recommended as a base for these preparations were yarrow (502, *Achillea millifolium*), chamomile (503, *Matricaria chamomilla*), nettle (504, *Urtica dioica*), oak bark (505, *Quercus robar*), dandelion (506, *Taraxacum officinale*) and valerian (507, *Valeriana officinalis*). Each of them has a particular characteristic, an affinity for sulfur, potassium, calcium, phosphorus, etc. Cultivating large amounts of these plants for use in compost would not be recommended because wild plants, when cultivated massively, lose some of their strength. It is therefore appropriate to reinforce their profile, their quality and to raise them toward a more powerful reality to concentrate their action. Rudolf Steiner recommended putting certain of these plants into an animal organ in order to, in a way, impregnate them. Of course, for each plant one must select the appropriate animal and appropriate organ from that animal. An organ that has an "affinity" with the plant and a plant that is "attracted" to that organ. To take an example, chamomile works most strongly on the intestines, a fact any herbalist can confirm. Chamomile's medicinal qualities are most pronounced in the intestines. Since for some reason there is a special relation between that plant and this organism, we will use it — if you prefer, we can say we are going to use the synergy between this specific plant and the organ on which it acts. Thus one must pass chamomile through an intestine, but which animal's? Clearly it should be the cow with its distinctive intestine, the longest of all the vertebrates. One could say that the cow is the animal which incarnates the best the digestive forces. Look at its fabulous ability to give milk. Burying these materials together further amplifies the relationship between the two substances. Autumn is the season of a descent into the earth. That which has lived above the soil will eventually be beneath it. Farmers in the past knew how, during winter, to find in the soil the effects of summer sun. For example, farmers often preserved certain vegetables in soil better than anywhere else. Preparations to be applied to compost should be prepared or completed in the beginning of autumn so that they can mature in autumn soil and then winter soil.

To work with these special affinities requires a pattern of thought that helps us to understand them. First of all, we must recognize that there is nothing mysterious about these traditional agricultural practices. In our time, with its lack of profound truths, mankind has a great need for extraordinary, magical things. We must take the opposite path, through trying to render seemingly mysterious things understandable for all.

Many ancient practices and remedies were abandoned over time without ever having been analyzed. These practices didn't involve magic, but rather an ability to treat something through an understanding of what is behind the effects we see on the surface. The effectiveness of many of these traditional practices have been confirmed by modern scientific methods, for example in studies made of traditional medicine in remote African villages. The quality of a preparation is a synthesis of all the degrees of these mutations. Their realization demands attentive and scrupulous care. For in the marriage of a plant with an animal sheath, the essential resides in the link that is created. This complementarity sheds light on what they have in common.

Another example is oak bark which is enclosed in the skull of a domestic animal and buried, always at the beginning of autumn. The bark is a process of rejection toward the mineral world while remaining at a stage that is higher and more alive. Bark is what the tree rejects. This downward movement toward hardness allows a link to be established with the bones. What these two materials have in common, though in different forms, is calcium. Oak is also marked by this tendency toward a hardness that slows down and ties into knots vital movements of growth. This tendency expresses itself through the rambling, agitated and slightly chaotic aspect of the tree. This preparation combines two types of minerals and a receptacle of spherical form.

A spherical form (like a barrel or an egg) is always a shape that draws in forces — the outside comes inside. We could perhaps say the macrocosm is becoming a microcosm. This is why the sphere has been so frequently used in architecture, with specific proportions or what is called the "golden proportion" no longer being understood. It was called "golden" because when you divide something, the parts of the whole do not typically bear all the qualities of the whole; this is a law of incarnation, and behind it lies the knowledge of the symbolism linked to Figures 1-9. With one specific figure (the golden proportion), however, the parts may keep all these properties, leading to its popularity in architecture.

Such forms, then, generate specific processes. Leonardo de Vinci pointed out some golden proportions in the human body. In our case, even if these proportions are not reached, sphericity brings something to the preparation. Thus, this oak bark preparation limits an excessive growth — which often attracts fungal disease — by helping to balance the two opposite stages that a plant must experience each year: creating its physical body (or moving into space, if you prefer) and returning, or disappearing into a seed. For the vine, this latter stage is achieved through flowering and making fruit.

The nettle is a valuable plant. It is capable of being immensely helpful to winegrowers. Nettle is concentrated in its roots which do not sink deep into the earth, but on the contrary remain on the surface like rhizomes, and its flower. This flower resembles foliage and never reaches the degree of dematerialization of a flower in full bloom. This detail explains the intensity of its action on the sap, so strongly dependent on the leaves. It is very effective in times of drought and is also a remedy for sclerosis. It is sometimes an antidote for systemic treatment products which are quickly absorbed into the sap of the plant and, therefore, interfere with the metabolism of the plant. It also regulates iron. Nettle is really the archetype of the leafy plant and in its living space it harmonizes all sorts of imbalances. It acts like a heart, an organ on which it has healing capacities. In the same way, the heart itself is permanently recreating a rhythmic balance between the brain and the metabolism. Nettle brings about harmony and can thus limit the invasions of parasites, which are sometimes necessary to regulate an excess of sap or some other imbalance. As a biodynamic preparation, nettle is buried in the earth for one year, to be charged with the energies of the four seasons of the year through the soil. It is one of the two preparations that does not require the use of an animal organ. It is the only plant that Steiner called irreplaceable — if one couldn't find it domestically, then it would have to be imported. One then mixes tiny quantities of the humus of nettle with the compost, in the same manner as the other preparations, using very small dosages. On the organic level, it can be applied without risk to vines as a tea and repeated several times as a remedy for the consequences of excessive sunlight. One can also spread it on the land as liquid fertilizer in autumn, so long as you never let it stink, because the earth is not made for bad odors. The only inconvenience of the nettle is that it does not cost money, which discourages all scientific attempts to confirm its multiple beneficial properties. How do you get a research grant to study a plant that is free?

Yarrow, on the other hand, is buried in the bladder of a stag. Why a bladder, why a stag ? A deer has an enormous sensibility. Through its antlers — the opposite of the horns of a cow, which are much more empty inside and prevent the inside from going outward — it is intensely linked to the outside world. A deer perceives and feels any noise or tension in the forest. A problem in its environment becomes the deer's problem. During the very large tempest that destroyed a significant part of French forests three years ago, gamekeepers found several deer that were dead but showed no sign of wounds. They were killed by their sensitivity.

How do deer receive this heightened sensibility? The capacity starts with the antlers, and it reaches the kidneys and the urinary system. All of us know how intense emotion in a young child can cause loss of bladder control. Putting yarrow in the bladder of stag shows a fabulous understanding of the subtleties or synergies that some animal-and-plant combinations can offer. We lose sight of these connections if we become lost in the materialism of modern science with its microscopes, etc.

With the yarrow preparation, we harness an enormous process of sensibility, or, if you prefer, a process of "listening" or of "welcoming," which is then given to the soil so that the macrocosmic forces — which carry the archetypal forces of each plant — can properly manifest. Our vines need these forces for building up perfect grapes. All these synergies can be used freely and offer a possible means of freeing our farming practices from the dangerous assistance of chemical companies.

Putting dandelion in the mesentery of a cow is also fairly easy to explain: a dandelion has the capacity to make one very strong, thick, downward-going root that is strongly linked to gravitational forces, while on the other hand it has the surprising ability to create a seed which is virtually weightless and can fly in the air. These properties indicate a capacity for linking the earth and the cosmic forces in a very specific way.

Now look at a mesentery: it is a very thin, bright skin around the intestines. It acts as a cosmic matrix — it could be compared to a sphere or to the shape of an egg. Thus, by putting a dandelion in the mesentery of a cow, one is reinforcing the descent into earth of a process of silica. In terms of current technology, we could call silica one of the most important "portable phones" to connect the action of the cosmos with the earth.

Valerian is easy to understand. Just look at it: it grows upward. Instead of flowering like most flowers do, when it is half a meter high it keeps growing and growing upward, at times as much as 1.5 meters. Valerian links itself intensively to heat, and it meets its archetypal forces by rising. This is the reason why it has so much phosphorous inside.

Once one understands this, it is clear why chemicals that limit the upward growth of wheat to 50 centimeters instead of 1.4 meters, as was the case 60 years ago, are so destructive for the intimate qualities this grain is supposed to bring to human beings. These upward-rising processes are always a call to qualitative cosmic forces which draw to earth a specific taste, smell, structure, etc. — all the qualities that are so significant for winegrowers. The originality and the success of their wines lie there. Only through this kind of knowledge can he abandon the technology that manufactures a taste anyone in the world can copy.

To summarize, different plants are enclosed in the organs of certain animals and buried in the earth at precise moments, in precise locations. At the conclusion of a period of ripening, they are dug up from the soil and preserved under rigorous conditions before being mixed in very small dosages with the manure the following autumn. These preparations can augment the bacteriological life of the compost by 30 percent or more and develop valuable processes involving calcium, potassium, silica, etc.

As a general rule, the vine needs very little compost. The quality is much more important than the quantity for, thanks to the preparations, the compost has acquired a totally new aptitude to generate life. It can then permit significant changes. Alex Podolinsky, who later developed over four million acres in Australia biodynamically, evokes in his two books, *Biodynamics: Agriculture* and *Introductory Lectures*, the very rapid transformation of chemical residues he obtained. From the earliest years of biodynamics, on soils that had been saturated with DDT, the harvest was healthy because the roots nourished themselves exclusively on regenerated soil. Six years later, the DDT had completely broken down into by-products that were less dangerous. This phenomenon is really specific to biodynamics and was brought about by preparations of quality.

Compost is a listening post that one gives back to the soil, an organism generating life, which one puts back to work by reinforcing the specific characteristics of the place of origin. We should work to build up a soil, to animate it, to make it more supple and expose it to the air. Let's get back into the habit of touching it, of holding it in our hands, of smelling it.

Living soil is always a delight to see. Be careful not to make it too rich by bringing in excessive amounts of powerful materials. That would be a major mistake. Some winegrowers, newly converted to biological agriculture, often fall into this trap. They yearn to replace what they have destroyed. Vineyard soil should be alive but rather poor, otherwise the quality of the wine suffers. The vine is made for struggle. The vine should somehow express this conflict in its grapes. For those who cannot use compost, Maria Thun

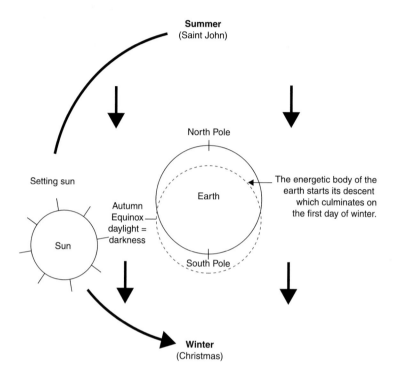

**The forces of internalization
of the setting sun
(Northern Hemisphere)**

Summer
(Saint John)

North Pole

Setting sun

Earth

The energetic body of the
earth starts its descent
which culminates on
the first day of winter.

Autumn
Equinox
daylight =
darkness

Sun

South Pole

Winter
(Christmas)

*For the Northern Hemisphere, the "energetic" body of the Earth, in the image
of the sun, begins its descent at the beginning of summer. Vegetation begins to
withdraw into itself. External growth stops, internalization toward the seed, the
growth of the grape starts with the beginning of the cycle. When the time for
plowing comes, the sap from the leaves descends to the roots. Our activities during the summer descend into the earth. This entire period is an inspiration for
the Earth.*

has developed a replacement product that is quite similar and
much easier to handle, a sort of mini-compost.

Winter Rest

One must try to understand this winter period, the meaning of
this life-saving repose for our soils and vines. Everything seems
dead and extinguished as the earth goes hard after a freeze. Full

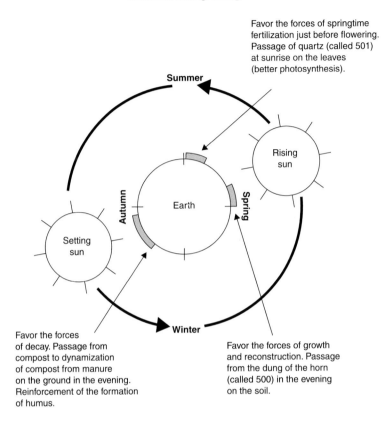

Principle moments of biodynamic action in wine growing

Favor the forces of springtime fertilization just before flowering. Passage of quartz (called 501) at sunrise on the leaves (better photosynthesis).

Summer

Rising sun

Autumn

Earth

Spring

Setting sun

Winter

Favor the forces of decay. Passage from compost to dynamization of compost from manure on the ground in the evening. Reinforcement of the formation of humus.

Favor the forces of growth and reconstruction. Passage from the dung of the horn (called 500) in the evening on the soil.

moons are high since the sun is lower. Death, however, is only in appearance. The exchange goes further and is more subtle. In the great balance, the primary forces which cause all growth on earth come down into the other hemisphere. They create an inhalation toward the South Pole. We are capable of feeling this attraction the earth has for the sun. When it crosses the ecliptic and passes above the sun, it is winter for the Northern Hemisphere and the days are shorter. The "vital body" of the earth, attracted by the sun, descends and moves off center toward the South Pole. For the northern half of earth, it really is a question of inhaling what is on top of the soil inside where the separated elements no longer have, at least temporarily, physical tasks to accomplish. Before the mobilization of spring, earth is at rest and waiting. At this time of winter the earth crystallizes, that is to say, it makes tiny crystals. One could say in the form of images, that what it hears in this calm generates a crystalline tendency, a form of purity and transparency. It is in tune with cosmic forces.

Winter is a descent, everything contracts into the roots of the plant. The vine and its soil receive the entirety of our agricultural activity during past seasons. If they have been reflected on and meditated over, we will perceive anew the benefits in spring.

Winter manifests itself in different ways. We should be sensitive to all these signs. The old-time peasants considered snow a gift from the sky, like a powerful fertilizer at a time when the soil was alive and seeds were healthy. Through the diverse forms of its crystals, snow manifests the power of the formative forces that one seeks so often in biodynamics. Children are still fascinated by this spectacle.

According to Rudolf Steiner, during this winter period, from mid-January to mid-February, man has the possibility of regulating the diseases of his estate by the strength of his thoughts.

At first glance, these statements seem absurd. Some will smile. However, many experiments have demonstrated that a grower can influence the growth of a plant without touching it, merely by his presence. Music also has an influence on the world of vegetation. These reactions are perfectly explained when one knows the intense sensitivity of a plant, of a vine and of an animal. During the winter the plant is not occupied with any material task and it is even more particularly disposed to perceive these influences. Thought is not measurable; it really belongs to this same qualitative and intangible world. Certain monastic orders have left the high places where, through solar meditation, they forget their terrestrial condition. They installed themselves at the bottom of the valleys and to fecundate the earth, which they work through their actions and their thoughts. This is why Cistercian monks began planting vines in France in the 12th century. It was a painful choice, but certainly well-founded. We must keep in mind that thoughts are also frequencies or microfrequencies. Without frequencies we could not communicate over long distances —

phones, radar and GPS would be useless without frequencies. Why should we deny their reality in our thought processes?

So long as it has not hardened, an artist can modify the contours of a clay sculpture. In the same way, so long as the springtime has not materialized an imbalance in the form of a visible sickness, corrections can be made more easily. The future vintage is in preparation during the winter. That is why the winegrower must be attentive to what his soil and his vines are trying to tell him.

The new medical tests give us an idea of the subtlety of a sickness at its beginning. Sensitive crystallizations consist of crystallizing a product to estimate its vital force. The image obtained somewhat resembles the stars which form on a windshield on frosty days. Human blood has been tested in this fashion. It turns out that all diseases become visible in a crystallization before the first physical symptoms appear.

Beyond a simple number of globules, the blood bears the mark of all our tendencies. An imbalance of vital forces in the blood is a potential sickness which has not yet manifested in physical form. These tests, extensively developed in Germany, remind us that a sickness passes through a subtle stage before emerging on the physical level. This is not surprising, since it is life which generates matter and not matter which generates life. This also applies to the world of vegetation. Each of us is free to accept or reject the action of thought on the health of vegetation. Wisdom dictates that we always keep a door open on the "unknown."

Chapter IV
Spring, the Season of Reconstruction

It may be that the earth was fertilized by cosmic forces during the course of winter. In the spring it gives birth to a new wave of vegetation, full of tenderness, refreshing and youthful. Earth has again crossed the ecliptic, the equinox has taken place, the days are getting longer, and the vines need our help to fully assimilate this new rising current which escapes from the soil. The vine must unite with this current in order to feed on it. A living and receptive soil can help the vine to reach its goal. It is the opposite of the season for compost when everything needs to break down. In the springtime, what has been separated should be brought together. Vine shoots, leaves, flowers and fruit will be born from this union.

The preparation necessary for this task was elaborated during the autumn. It ripened underneath the earth during the winter in a cow's horn that had been previously emptied of its cartilage and then filled with dung. Dung, horn and hibernation — three essential components. Dung we already know about. The vegetation absorbed into the intensely active digestive system of a cow has acquired very important forces of growth. The feminine aspect is equally primordial. Our terrains ("terroirs") should constitute a living organism of greeting and reception. This is why former civilizations spoke of "Mother Earth." By the choice of a cow's horn (not a bull's, with which results are almost nonexistent), we are going to accentuate this femininity. But why put the dung in the horn?

The Meaning of the Horn

A herd of dehorned cows (what a wretched custom) carry their heads as if they were too heavy. Sometimes their heads almost touch the earth, they are so low and heavy. A nice pair of horns in the form of a lyre seem to make a cow's head lighter. It adds a touch of pride, of nobility, of highness. Unfortunately, these qualities are "without weight," that is, they are no longer considered significant in our era. In the mountains, where the most powerful solar forces have not yet been overshadowed by certain practices, where the flowers are more colorful and the honey is tastier, it is still possible to contemplate cows that carry their heads held high. By the way, this lightness echoes a physical detail, since the cartilage of these horns, formed of quite numerous cavities, is also more aerated.

The hollowness of a cow's horn is designed to keep inside the animal energy currents that would otherwise be given off. A cow is the opposite of a deer, and a horn is the opposite of an antler, which acts as a means for perceiving the outside world. A cow is completely "listening" to what happens inside itself. Look at a cow ruminating: even with the noise of a violent car crash, it will slowly turn its head toward that strange noise, without stopping its chewing. A deer would be 100 meters away within two seconds. On the one hand, the cow's horn receives sun forces from the outside, but the horns (and the hooves) primarily permit strong digestive forces to remain inside the cow; they prevent the outside physical world from acting too strongly inside the animal.

By placing these observations in the context of our four states of matter, it is easier to understand the meaning. The health of an animal can be evaluated from the way it carries its head and the shape of its horns. One day while I was accompanying Mr. Pouencet, who had been trained by veterinary Dr. Sellinger in Austria, I heard him describe from a distance — just by looking briefly at the animal — its character, its nervosity, its tendency toward mastitis, etc. The proprietor, who was present at the scene, could only confirm his analysis. And what if that turned out to be the work method of the veterinary of tomorrow? The use of antibiotics cannot go on forever. This quick fix sometimes heals conditions otherwise impossible to heal. It is not a question of miracles, but of understanding forgotten backgrounds and their secret actions. Extinguishing the vital forces of a farm animal with excessive food that generates fatty meat is not an end in itself. This practice will have to stop one day.

Each animal needs its vital space and its instinct, which is the protector of the quality of its race and helps it to be respected. The grower, torn between the needs of the species and the laws of the market place, usually chooses the path of profit. The cows (or other domestic animals) crowded into minimal spaces wind up attacking each other. They are dehorned to avoid accidents. No one seems to want to understand that by fighting or chasing other cows away, a cow is trying to communicate that it lacks the space it needs to remain in good health. In doing so it is being faithful to its archetypal forces, which work to ensure the species does not lose its qualities. Several generations later, the breed deprived of space and the solar aspiration that the horns bestow, no longer really conforms to its species and becomes more vulnerable to disease, leaving it more in need of help.

Only in our era can the horn be assimilated to dead matter, without worrying about the blood which gushes when one dehorns a cow. Our era has also permitted certain specialists to affirm that it is entirely possible to replace a mother's voice to a nursing infant with a recording. The nursing infant is not supposed to know the difference.

Man has not always lived in such ignorance. The profound meaning of the horn marks the pages of history. Is it not surprising to see the reproductions of ancient Viking warrior helmets, from which rise imposing horns? Would the idea come to you, if you had to fight for your life, to put such an arsenal on your head? What was being referred to, a symbol or a feeling? The two are often confused.

When the Egyptians placed lyre-shaped horns on the heads of some of their goddesses, their gesture had a completely different meaning from that of today. It was the solar forces which were thus suggested, if not precisely designated, as in a statue exposed in the museum of Luxor, where the sun is placed between the horns of the animal.

During the Middle Ages at the tables of the nobility, meals were announced by saying, "The water is horned." Water was brought to the table in horns to charge it with vitality. Everyone knows the fertilizing properties of powdered horn, which is sold as fertilizer. These qualities are unique and specific to the organ. Even the skin that surrounds the horn is not gifted with the same qualities. And the flood of wealth that streams forth from the horn of plenty? Is that illustration merely fortuitous? There is without any doubt a knowledge to be drawn from all these past witnesses — witnesses who support each other — to once again find the meaning of the horn.

Rather than using the powdered horn as fertilizer, biodynamics uses the horn as a container and fills its dung with formative forces (read *Agriculture of Tomorrow* by Dr. Eugen Kolisko, who was able to bring into relief their particular effects). We will not deal here with the making of this preparation, but all the details are important: quality of the horn, quality of the dung, choice of a place for its hibernation.

When one intensifies what is alive, everything counts. Expressions of the terrestrial, aquatic, igneous and aerial states should be present. A horn can be used up to three times; until the moment when it starts to flake away, as if to underline that its formative forces are no longer expressing themselves.

The horns are buried at the autumn equinox. After the spring equinox, they are dug up out of the earth. The appearance of the dung has been transformed. To convince oneself, it is sufficient to place some of the same dung in, for example, an earthenware pot, and to compare the results. The consistency and the odor of the dung from the horn is totally changed. Its bacterial activity can be up to 80 times superior to that of the test sample in the earthenware pot.

Is this not an ecological way of using the powers and specific qualities of nature? It is urgent to make the mentalities of those who preside over the future of science evolve. But this evolution is not possible without profound modification of school and university education.

The domain of the vital forces which animate our planet is absent from what is taught in the school curriculum. This concealment is a form of castration. Very few individuals are capable of overcoming such a handicap. The global system of life is dissimulated. It is thus "permissible" to violate it by imposing through genetics and other methods elements which have no place in the world of vegetation. The field of genetics is interesting, but modifying genes without understanding the fantastic system that has put them in place is a cosmic mistake. One should work with this system, not blindly against it. Progress does not come from such forced marriages, but on the contrary from the understanding of the complementarity and harmony of organisms.

Dynamization

The contents of one horn is sufficient for one hectare (2.471 acres). This treatment can be repeated two or three times in the springtime on dates called "of heat" or eventually "of light" (see Chapter VIII). The dung of the horn is dynamized in water, preferably luke-warm, during a precisely selected hour. Dynamization is an operation that is essential to the effectiveness of the preparation. It consists of stirring water in a circular receptacle, vigorously in one direction to create a deep vortex, then to suddenly reverse the direction of rotation (see illustration below).

A man's hand is important at the moment of dynamization. It is preferable to carry out this operation by hand. When one has to use a stirring machine, it has to conform to many precise criteria and one should refer to the deep work already mentioned by Alex Podolinsky. Putting water into movement in a rhythmical manner has a significance. Civilizations of antiquity, particularly the Celts, often made references to it. Their sculptures are often marked with designs of the creative vortices that increasingly play with

our atmosphere. Between the laws of condensation and evaporation, between the cold and hot zones, air is increasingly taken into these spirals which intensify all sorts of exchanges. One finds them further down the stream and in the ocean waves. Theodore Schwenk's book, *Sensitive Chaos* is illustrated with numerous photographs in which he captured the images of these creative surges which unceasingly reproduce the wings of birds, the fins of fish, etc.

At this point each living creature is an animation of the vast earth organism. For example, the human voice at each note modulates the air into multiple forms. Have you tried to compare the effects of mineral water fresh from its spring with mineral water from bottles? Try the experience at Vichy with a little glass of "living water," dynamized by the movements of the spring. If you exceed prescribed dosages, you will be rapidly brought back to earth with a thump. There is an effect which a whole bottle of the same water would never provoke. Dynamization creates a profound intimacy between the solid and the subtle, between what one dynamizes and the water. The rhythm acts like an appeal. The product is filtered at the end of an hour and is passed through a small sprayer that produces large droplets, or is sprinkled with a little broom to spread the drops of fertile liquid on the soil.

It is of course an evening treatment, since its focus is the soil. It can be applied until dusk. Outdoor temperature should not be less than 45 degrees Fahrenheit. Cold, which is linked with death, is not compatible with fertility. One should consider this before proceeding with artificial insemination that reduces the act of reproduction to the union of an ovule with a sperm.

Even on the stud farms, the stallions are being replaced by test-tubes. Haven't you seen a bull isolate the cow he wants from the rest of the herd before mounting her? He keeps her isolated for several days and remains near her. That is another reality that expresses itself, an intangible reality, thus disconnected from the contemporary mentality. But those who have remained sensitive can understand why and how the methods of artificial insemination weaken a breed from generation to generation. Research tries to remedy these weaknesses yet again through genetics. What a lack of balance, and what prospective profits.

in an era when immense energetic and physical pollutions are slowing down the process. The agrichemicals act in the same way by obstructing the pores of the leaves. Today photosynthesis, through lack of understanding, is unfortunately perceived as an indestructible automatism. How ignorant.

This preparation is a remedy against the disorders of the atmosphere. It is going to help the light and heat descend and become more intense. It favors the making of fat and aromas. It reinforces color, and makes the structure of the wine more dense. By approaching the solar forces, this preparation touches the vine in its entire being. When a vintage ripened to maturity encounters the effect that quartz has on the wine-press, the sieve is sometimes almost clogged by this material condensed as "oil."

The use of this preparation, the 501, is simple. A quarter teaspoonful per acre is sufficient. Rudolf Steiner, at the time of his conference in 1924, suggested dosages that were reduced even further. Since then, certain imbalances have been amplified. The 501 is dynamized in water for an hour. Being intended for leaves, the application should be made very early in the morning, never later than 9:00 a.m. Sunrise is the moment when the effects of synergy between the quartz and the rising sap are at their peak. At noon they become destructive. Dynamization is performed for one hour, as for preparation 500 (horn manure). If one performs such a treatment at the moment of flowering, the excessive light generated carries the vine along to the next stage. The risk of run-off is then much greater. It is thus a good idea to give oneself a margin of security of several days before the flowering. On the other hand, if treated too early in the season, it can diminish the growth to the extent that the process of flowering takes on an opposition to the forces of terrestrial growth, which would then undergo metamorphosis. It is preferable to spare the buds, which have very little surface to absorb it.

In my opinion, this treatment is too often misused. The basic rule to keep in mind is that 501 will increase the work that the vine is doing. In spring, therefore, it will force the vine to make more leaves and wood. Often we simply destroy the extra work we have forced on the vines two or three weeks later, when they threaten to become a messy bush. Instead, one might consider spraying it high enough every seven or eight rows just to illuminate the air — this is what it does; it improves the quality of light, which will then be used for photosynthesis without bringing too much extra work. The effect of such "spring spreading" may be seen in more pronounced colors of leaves in autumn.

Be sure it does not fall on naked soil, as it will destroy soil life to a depth of a few inches. It is the reverse of 500, its opposite polarity. When applied to plants, on the other hand, it is a fast way to develop soil life. If there is a cover crop between rows, for example, this preparation is very useful because it will increase bacteriological activity around the roots. Its effect is the opposite of a weed-killer, which puts poison deep into the soil through the roots of the plant treated. With 501, the roots instead bring life forces to a certain depth. Keep in mind that this treatment will always force vines to expire water. If you have an irrigating system — which is not recommended, as it slows the descent of roots into the soil, thus counteracting the Dionysian process — you can control this aspect. After a strong storm, 501 is a gift that can help get rid of excess water and also suppress fungus disease.

Two or three weeks before the crop appears, spraying 501 before or at dawn will result in fabulous taste. The 501 preparation helps the vine perform its duty, giving the grapes all those qualities we like to find in a wine. If 501 is spread too close to (within two or three days) or during fruiting, it will kill your yeasts, and fermentation will last much longer and have a difficult start — but by the same token it can really repair the damage of an excessive rain during fruiting.

After the harvest, some winegrowers renew this treatment on the leaves (or on the soil), in the evening when the sap retreats. If the treatments at the end of spring have been carried out correctly, this precaution is undoubtedly not necessary. The quality of light is not that of the solstice and, in any case, our spring and summer activities go back down to the roots during this period of the year. Under these circumstances, such a treatment should not be generalized, but considered as therapeutically adapted to a particular situation.

Resorting to 501 is tempting for those who are not practicing biodynamics and thus have not converted their land or even applied preparations. It is necessary to warn them. The effects of these treatments are above all linked to their complementarities. Used on its own, quartz can become dangerous, for the excess of light generated should be balanced by a good strong soil. This balance is necessary for the expression of life. Biodynamics supposes an understanding of the meaning and repercussions of our actions.

Complementary Actions

Beyond biodynamic's three great moments of action, which we have briefly discussed, certain complementary actions can turn out to be necessary during a period of reconversion. Intervening

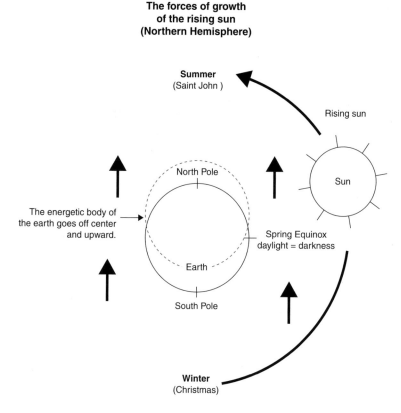

**The forces of growth
of the rising sun
(Northern Hemisphere)**

Summer
(Saint John)

Rising sun

North Pole

Sun

The energetic body of
the earth goes off center
and upward.

Spring Equinox
daylight = darkness

Earth

South Pole

Winter
(Christmas)

Since the beginning of winter, in the image of the sun, the "energetic" body of the earth begins its ascension. It is the beginning of an "attraction" of vegetation upwards. After the spring equinox, we witness an externalization. The vine forms its budding branches and leaves. During this season, the earth exhales.

on a biological domain is entirely different from intervening on a domain that has been mistreated for years with herbicides. It is thus possible to consider a "cleansing of the location," not by a disinfection, about which we know the qualitative effects, but in making use of homeopathy (see Chapter VI). A high dilution of thuja (cedar) or of calendula (*Calendula officinalis* or pot marigold), that savage little marigold intense with luminous forces, is sometimes a great help. Even when uprooted, for several days the plant still has the strength to turn its leaves toward the sun, thus accentuating the intensity of this link. These dilutions, spread on the soil preferably in autumn, are capable of awakening the vital forces and favoring the expression of biodynamic preparations.

To Make an Herbal Tea

In the following order, one could use a tea made from chamomile, yarrow, nettle, oak or dandelion. Many others plants can also be used. For one acre, take the equivalent of a handful of the flowers. When the water boils, stop the fire and insert the flowers. For nettle tea, it is sufficient to cut an armful of nettle with the stalks and to repeat the same operation or to put them in cold water and stop the fire when the water comes to a boil. An armful in our era when weight is so important consists of two hands holding the stalks together. For oak, it consists of a brew. One breaks the bark up into little pieces and boils it for 15 to 20 minutes. The tea can be diluted on a basis of 9:1, that is to say, 10 percent tea to 90 percent water. About 8 gallons per acre are necessary. More can be used, if one wishes, but it is the quality of the flowers that counts.

For a decoction of oak tea, one should use the outer bark, from a tree that has recently been cut down or is still standing. Nettle remains without a doubt the most precious of the teas for its mediating properties, its ability to harmonize. It can be applied several times, to the leaves as tea and to the soil as liquid manure. In summer, a bath of liquid two days old is sufficient. It should be kept a bit longer in the autumn, when the weather turns cool. Nettle is what supports the leaf. This tendency is accentuated by a maceration of 12 hours, water being, among the four states of matter, in correlation with the leaf. On the other hand, the maceration of floral teas should be brief, no longer than one hour.

Herbal teas also belong to the category of additional practices. We will first mention the plants designated by Rudolf Steiner for the making of preparations — chamomile, yarrow, nettle, oak, dandelion and valerian. The effects will obviously be quite different; there will not be any transmutation of substances, but they will be perceived directly by the vine, for they should not pass through the earth. Certain plants with an affinity to the vine can be beneficial to it. It is also possible to follow the opposite line of thought and bring it a tea that upsets it a little. We know its force of reaction. It would be interesting to then feel its effects. Herbal tea acts immediately, that is its main interest. It is put on the leaf and passes directly into contact with the sap. It permits the vine to generate life in its environment. And, while awaiting the rebuilding of the soil, herbal tea is a useful support because, in contrast to foliar feeding, it brings a tendency, a movement, rather than dead and stillborn elements. The odor of these teas is already an encouragement. By smelling them, the winegrower has the impression of feeling their benefits. That never happens with agrichemical products.

Protecting the vine from heat can be achieved in several ways: through altitude, or planting on north slopes or in places with cool nights. Teas made from specific plants can also be used. One group of plants that are useful in cases of aggressive sunlight are some common seaweeds — *Fucus vesiculosus, Laminaria saccharina* or *hyperborea* or *digitata.* Salt produces a terrible burning process that will kill most plants. What does seaweed have that enables it to resist the salt of the sea? It surrounds itself with colloids — exactly the substance we need to spray on our vines against an aggressive sun.

To make this preparation, fill a garbage container (60 liters is more than enough) with seaweed and submerge in water once or twice for a few minutes. Cut the seaweed into small pieces of 1 or 2 inches, then soak them in hot water for three to 12 hours (or longer, if you desire). This tea is very efficient against sunburn — it is used in many creams — and it is also beneficial to the vine. On America's West Coast, where the sun is really harsh, I would also be inclined to also use aloe, a member of the lily (liliacés) family, which has enormous properties against all burning processes. The Yucca and eventually a cactus could also be helpful, but to a lesser extent. A tea could be simply made out of crushed branches. It would be interesting to measure whether it reduces the level of alcohol.

Very little is needed. Just as the quantity of active ingredients in medication is tiny, so it is with agricultural teas. A 50-liter garbage container full of seaweed will produce 300 liters of tea, and that can treat 10 or 15 hectares. In general, most plants can be mixed together to make a tea. The more plants in the tea, the smaller quantity of each you need — 30 to 40 liters per hectare of such tea is enough. Spray it in the morning or in the evening, but not in full heat.

The willow shows in spring its capacity to make buds that attract bees. Bees, of course, are attracted to pollen, which means that the willow has the capacity of linking the water level — remember it can evaporate 4,000 liters of water per day — to the light level of the pollens. This can sometimes be helpful for our vines.

The books of Grohmann and Pelikan are full of information of that sort. In my mind, they are by far the best books ever written on plants. They have linked the knowledge of the past to the knowledge of our times, which focuses on the physical level.

The more we understand the "behavior" of each plant in the way Goethe explained it, the more we realize that we have around us a number of possible teas to help our vines perform their work well, for example, in balancing some craziness of the climate. Heat plants can help against frost (thyme, sage, cloves, valerian). A tea of rosehips applied six weeks before fruiting can help the vine to start a sugar process in its grapes. Willows, nettle and seaweed can help resist an excess of sun, as discussed above. Nettle stimulates the circulation of the sap, but be careful with it if you are worried about high alcohol content in your wine — on a dry year, a nettle tea can produce one extra degree of alcohol!

Garlic can bring back life forces and health, but don't use it too close to fruiting. Once more, keep in mind that this will work well only when your land is alive; on land that has been poisoned by weed-killers, systemics, etc., these methods will not achieve their potential.

The Harvesting of Plants

As a general rule, plants are picked at sunrise. Morning is a privileged time. The appearance of the sun is an event for the world of vegetation. It captures its attention to the highest degree. It is thus a significant hour for harvest, even if it is still possible to slightly prolong this moment, so long as one does not fall under the first rays of the noonday sun.

It is obviously preferable to pick our plants ourselves, in a place that looks healthy, avoiding the sides of the roads. It is the feeling of the plant and not just the type that interests us. A herbalist is not always able to respond to such demands. If you follow the order in which the teas are given, yarrow is harvested for the following year, for it had not yet flowered at the propitious moment. It is sufficient to dry it in an attic and to shelter it from light in a box. Don't hesitate to take a little yourself.

Some people dynamize the teas or make homeopathic dilutions of them, but this may not be necessary.

Under the Summer Sun

We hardly speak at all of summer since, in principle, during this season there are not any activities directly linked to biodynamics in wine growing. The grapes are formed and one can plow back earth on the feet of the vines. The soil should remain loose so it breathes well, without being dried up by excessive hoeing. If the dates assigned to this work correspond to the "heat" influence (see Chapter VIII), that is a complementary factor in the quality of the wine.

Faced with the possibility of climatic extremes, it is possible to resort to certain preparations or teas to try to re-establish harmony. Against excessive sunlight, one can think of nettle tea or a tea made from mulberry leaves — a tree which allows the silkworm to link itself with light to become a butterfly. In regions where summer is almost too violent for the vines, it may be useful to spray the leaves with an atomizer containing dilutions of claylike bentonite, or to make a "500" on the soil or eventually on the leaves, from dusk to late in the evening, which will generate an abundance of dew on the leaves. Against excessive rain, a dynamization of quartz (501) has a drying action through its affinity with light. If summer is too cold, we turn to valerian, which is associated with heat. The winegrower analyzes many situations case by case, keeping track of the location, the date, etc. In general, a vine raised biodynamically after several years has a highly increased capacity for reaction when faced with imbalances. Its sensitivity to remedies that we choose to bring is very definitely improved.

These comments may suggest that biodynamics is something really difficult, but it is not. One simply needs to apply treatment in the fall one or twice with the compost. In the spring, apply one or more treatments of 500 when the soil is ready. Also apply 501 in the fall or spring, or both. As you observe the results, you will be able to adjust applications for the following year to a treatment that is better tuned to your vineyard. This is the process that leads to a unique wine.

I have to say that I am personally not in favor of the "prepared 500" created in Australia a few years ago. It includes the preparations 502 to 507, and was probably intended to save time by avoiding separate applications. I see this as a significant misunderstanding of the biodynamic method. As Bernard Lievegoed explains so well in his booklet on biodynamic preparations, 500 and 501 are like two sides of the sun: one is the "cosmic sun," and the other is the "sun of the soil." In our solar system, the sun is like the conductor of an orchestra, and combining other preparations with 500 creates a significant problem in rhythm. This is especially true of the vineyard, which does not grow in the fall like wheat and which is so deeply linked to the four seasons. By regarding the soil alone, one cannot achieve a fully balanced system.

Using a prepared 500 could be compared to a chef of a good restaurant deciding to bring you the entry, the main course and the dessert mixed together, claiming that in any case it would be mixed in your stomach after absorption. Understanding of natural rhythms is essential for a deep understanding of biodynamics.

Chapter V
Sickness is a Sign of Imbalance

This chapter is not intended to deliver a dictionary of recipes. What is essential is to demonstrate the urgent necessity of treating diseases from a different point of view. Diseases terrorize the winegrower and cause genuine cases of psychosis since, in some cases, they destroy a harvest in two or three days. Fear never gives good advice for correct healing.

The basic rule that should be taught in all farming school is that for plants (it is not fully true for animals or human beings) diseases are simply a lack of life forces, that is, a loss of connection with the system that provides life and health. When we use synthetic supplements, the artificial molecules we force into our plants' sap disturb their link to the complex energy system that provides balanced life. The more we treat with these chemicals, the more we create an environment for side effects or "new diseases." We have to understand that nature has created a number of viruses, microbes, etc., whose duty is to destroy plants (and animals) that are lacking life forces. Their directive is to find that which is not in harmony with the rest of nature and destroy it so that it can be recycled.

Some new molecule from the laboratory may stop a specific disease, but it will also open a space for another one, since the *cause* of the first disease has not been addressed. The winners in such a system are the chemical companies, while the losers are the winegrowers and the consumers who eat foods that contain no harmony. In fact, this is the beginning of a cancer process: too much material or cellular forces and not enough structuring or cosmic forces. Industrial food has lost its connection with these higher forces. Today's dominant agricultural system may provide food to the world, but quality-wise this food is extremely poor.

With organic practices, a significant step forward is achieved with the banning of all unnatural products. This is great progress at the physical/material level (natural manure, cover crops, etc.), but biodynamics is designed to act at an energetic level — it recreates a link to the natural system of energies and forces. At this level, small quantities have big impacts — 10 to 100 grams of supplementation per hectare is enough.

In the realm of energy, weight is meaningless — the laws of Newton have little meaning when we leave the earthly level. In our paradigm, then, diseases are just a disconnection of the plant from the energy matrix of life, or they might be the result of "a good thing in the wrong place," an organism that has left its proper sphere. Ultimately, diseases have their source either in the denaturing of the atmosphere or in the destruction of the life of the soil.

Regarding the atmosphere, we have to be aware of energy pollution, which is never widely discussed for obvious economic reasons. Human beings are filling the atmosphere with all sorts of frequencies through satellites (with more than 2,000 currently in orbit, and more than 100 new ones added per year) radars, relay for mobile phone, etc. To picture how this affects the earth's atmosphere, imagine that you are talking to someone at the end of a long room. If 100 nervous people suddenly entered and began shouting, your conversation would certainly be disturbed.

With our communications and energy technologies, we are doing something similar to the earth's atmosphere, and with increasing intensity. Remember that the earth receives its information from the solar system through frequencies. Has anyone investigated how our systems of communication affect or interfere with the earth's communication with the solar system? A mobile phone uses gigahertz, that is, it produces many millions of vibrations per second. The atmosphere is a complex, sophisticated system that we disregard at our peril. Not only is our health affected, the earth itself, which is a living organism, cannot tolerate this pollution much longer.

Imagine living in a town with all these antennae, satellite dishes, cars with GPS — which means wavelengths following each car — microwaves, mobile phones, etc. How do these affect the frequencies that keep us alive? We need qualitative tests (see Chapter IX) to answer these questions. What we can do as winegrowers, however, is use the 501 preparation to restore our vines with the forces it needs from the atmosphere, or, if you prefer, it attracts and allows proper, healthy life forces to penetrate the energy pollution — like creating a special line of communication. In doing so, it increases the quality of photosynthesis.

Health Resources can be Built Up or Destroyed

Let's try to understand our actions as integrating into a vast organism which forms the four states of matter. Examples are numerous. A vine that has been forced to grow in winter during a period in which nature is not propitious for growth, under the effect of artificial heat, or doped with a variety of substances including hormones, ceases to be able to integrate harmoniously with its surroundings and reacts to them badly. A selection of clones, whose manner of birth implies an even greater isolation, would suffer from similar weaknesses. Yet for the last few decades, clones have represented 90 percent of the demand. From one generation to the next, these weaknesses accumulate. This simple fact also brings into question the selection of vine-stock and the reproduction of portable grafts. They are already injurious to the profound natural sensitivity of the vine. And what if, to put a stop to epidemics of new diseases, we begin by changing the vegetative material of which our vines are made?

Dates of planting and transplanting also influence the health of plants. In Germany, the Department of Water and Forests clearly reported that some stands of coniferous plants did not come down with diseases that nevertheless afflicted other stands that were in the same region. The first reaction of a technician would be to leap upon his microscope and undertake a whole series of analyses. The responses thus obtained are always superficial, for only a global approach is able to deliver real information. In this particular case, the reading of plantation records that were centuries old provided evidence that coniferous trees planted in the ascending phase of their planet — in this case, Saturn — develop increased resistance to disease. In other words, these trees were more strongly linked with the dominant aspect within them, which nourished them on a more subtle level (see Chapter VIII).

The respect of dates for cutting the vine-stock was essential during a period before fungicides and herbicides existed. If the soil is healthy, parasites do not attack vine-stock that has been cut and planted correctly. This phenomenon accentuates the positive aspect of diseases, on which parasites most frequently play a regulatory role. There is one basic rule in this domain. A planting of seeds should be "attracted" from above, thus put into the earth during the waxing phase of the moon, which in a way reproduces the "ascensional attraction" of springtime. On the other hand, a transplanting, which is always a shock for the roots, should be drawn to what is below. It is a question of recreating its link with the earth. A waning moon is favorable to this movement. This approach can be considerably refined if one takes into consideration the less frequent revolutions of the planets. There are periods which are particularly propitious for planting the precise essence of a tree. You will never obtain an oak thousands of years old without this knowledge. Oak is dominated by Mars, the vine by Mercury. This is not a matter of superstition. All the ancient works of botany refer to this knowledge. In Germany, Maria Thun carried out numerous experiments on this theme. It turns out that the choice of different dates for working the soil, accompanied by dynamizations, can change the form of fruit and leaves. The only thing we care about is allowing the plant to be harmoniously integrated with its surroundings and thus more resistant.

Reserves of Health are Built Up or Destroyed

Unfortunately we have reached a dangerous threshold in the destruction started several decades ago. The disorders thus generated *are* something to worry about. The vine is part of this reality and its reserves of strength have been considerably weakened.

Fungi Leave the Sphere
in Which their Presence is Beneficial

Cryptogamic diseases, like powdery mildew or oidium (*Uncinula necator*), are the most frequent. These diseases are linked with fungi, but the life of fungi is indispensable to the operation and expression of the soil. Hildegarde de Bingen described them as "the foam and sweat of the earth." When this fungi leaves its sphere of activity to climb in extrapolated form to the level of the leaf, its presence becomes dangerous. Why? For decades this question has been ignored. The solutions proposed always bring us back to two anti-cryptogamic treatments which, while suppressing the disease of the leaves, destroy all cryptogamic activity in the soil. The results are conclusive, if one doesn't take into consideration the trap in which the winegrower is caught. The more the soil is destroyed, the greater are the imbalances which result, which generate increased vulnerability to disease. Yeasts, marked with the climatic profile of the year, are also destroyed to such an extent that the necessity of re-yeasting is now part of current practice. Such are the yeasts of commerce.

Products for actual treatment most frequently belong to the family of systemic plants. They are absorbed into the sap during the hour which follows the spraying of the leaves. Rain can no longer wash the product off, consequently all the new shoots are protected. This method is certainly very practical. However this invention is at the origin of the qualitative drama. The plant is a whole entity and the sap supports its life. It is the sap which makes the grape. Today we know the secondary side effects of these treatments: weakened metabolism of the plants, residues (the time required to eliminate the poison varies according to weather conditions) and, sometimes, problems with the taste. The most serious threat was the action of systemics on the roots which should generate microflora to assimilate the soil. No anti-cryptogamic is made to help with this.

Before accepting the teachings of biodynamics, the winegrower should ask some perceptive questions. Why did this cryptogamic life-form, so useful at the surface of the soil, rise to the level of the leaves? How can we contain it within its sphere of action? Rudolf Steiner gave several answers, at a time when agriculture did not yet suffer too much from degeneration. According to him, when the moon is full and at perigee, and the soil is humid or it's raining, the forces of growth in the soil can become excessively potent. Indeed we know that earth is not exactly at the center of the orbit which the moon makes around it; the earth is slightly off-center. The result is that during each of its 28-day cycles, the moon passes through a far-off point, called the *apogee* and a point that is near, called the *perigee*. A full moon totally reflects the light of the sun toward the earth. If besides this, it is at perigee and thus near, it reaches the totality of its power. The moon, as is evident from the phenomena of tides, is strongly linked with water. Its action is thus reinforced when it rains. A mushroom can be thought of as a flower of earth since it opens toward the bottom, not the top. In the situation we are describing, the forces of earth mount above their physical limits and generate an inappropriate life in their near environment, in this case the leaves.

We described this phenomenon while discussing the specific qualities of vegetation near the equator. It is thus a question of limiting the excessive terrestrial force. Rudolf Steiner advises the use of horsetail from the fields. The main quality of horsetail is a very strong concentration of silica. Its ashes contain over 90 percent silica which, because it is in vegetable form, is close to the vine and easily assimilated. Let us remember that the forces of light and heat transmute what rises from the soil into a more subtle state that manifests as a flower, in which the earth forces are not as dominant. In the same way, by making a decoction of horsetail and spreading it on the soil, we restore balance by reinforcing the powers of light and heat. That is the principle.

Horsetail "fabricates" two stalks, one after the other. This is a characteristic we are going to make use of. The first stalk enables it to emit spores and accomplish its terrestrial task of reproduction. On the second stalk it makes its leaves, which look like needles, demonstrating well the influence of silica. This ability to separate its terrestrial function from its other nature is valuable in the struggle against cryptogamic disease.

It is above all the horsetail of the fields (*Equisitum arvense*) which concerns us. Horsetail from swamps or forests, in spite of a physical resemblance, is much less rich in silica which is certainly understandable. If you buy it in powder form verification is difficult, and even in good faith an error is possible. Then one must make sure it has not been exposed to gamma nuclear radiation, like most of the herbal teas sold in pharmacies. This procedure, illegal for food products in many countries, is proceeding full speed ahead in France. One day I had the opportunity of asking a government official why the term "irradiated" is not used (they prefer the term "ionized"). He pointed out to me — dead seriously — that products labeled "irradiated" would be impossible to sell. This infinitely pernicious treatment remains unspecified. This is completely illegal. Tests of sensitive crystallization on previously treated potatoes, provided evidence of a considerable modification in the creative life-bearing forces. But a society obsessed with criteria of quantity and profit does not pause over such details.

In any case, it would be useless to count on any effect from irradiated horsetail. It would be a waste of time. The location it came from is information that should not be neglected. The horsetail imported from the Eastern countries is often very polluted, if not radioactive. This is also the case for certain types of peat. Of course all these precautions are not always easy to take, but one might as well act with knowledge of the cause. The quantities required for a decoction of horsetail are quite small: an armful of the plant per half-acre or a double handful of the powder, left to boil for twenty minutes. The decoction is spread on the soil, instead of on the leaves, whose metabolism it slows by hindering the circulation of the sap, especially if the weather is hot and dry. It is a question of preventative treatment, from which you should not expect total protection. The duration of its effect may vary. Let us not forget that we are trying to bring two opposite forces into harmony. Renewing the treatment in the middle of June is sometimes suggested, but overdoing the treatment may endanger the life of the soil. Against mildew, the use of horsetail permits one to drastically decrease the dosages of Bordeaux copper sulfate, especially if the oatmeal is mixed with an infusion of nettles. Four or five treatments (on a basis of one to two kilos per hectare) are usually sufficient.

Horsetail is also effective against oidium, knowing that this disease does not pose the same problem, since the vine is treated with sulfur, a very beneficial luminous process at its moment of flowering. According to Rudolf Steiner, homeopathic dosages of sulfur are present everywhere in our atmosphere and permit the appearance of all forms of life. Putting sulfur on a vine a few days before its flowering helps the process of flowering. It is all a question of dosages. There is no point in spreading more than seven or eight kilos of flower of sulfur per hectare. Over 10 kilos and the acarian (beneficial mites), who are predators on yellow-and-red spiders, die off. Using sulfur in liquid form is less effective, being further removed from its deep nature. Be careful of the types of additives sold to make it soluble. Some of them are toxic and therefore not accepted in the program of biological agriculture. There is another efficient and cheap product against oidium: milk! We have used it here successfully for the past three years — eight to 10 liters by hectare, diluted in the amount of water you need. This should be unpasteurized milk, organic preferably. A good organic milk also seems to be beneficial to the vine and its leaves when the sun is too harsh, whereas sulfur may burn them, especially if the temperature is too high. Cost-wise, milk is about one-fourth that of a conventional treatment — perhaps that is the reason that no school talks about it?

There are other methods besides horsetail that act naturally against cryptogamic diseases. It makes one think of essential oils, of certain mints like lavender, which is strongly linked with heat, or of pine oil. In this domain we are at the kindergarten stage of knowledge. The field of research is wide open. This work should be carried out by a research institute, but we should remain prudent about interpreting the results. On test areas drenched in all kinds of toxic products for 10 to twenty years, the dosages needed to be particularly high before any improvement was noticed. On a domain where the laws of life express themselves, the situation is entirely different. I have been using sage *(Salvia officinalis)* against mildew with some success, a treatment that is beginning to be recognized by scientists. Onions or garlic are also good treatments, since they are very beneficial to the health of vines. In all of our preparations, we have to introduce a plant (in the form of a tea) with properties that are compatible with the vine. If not, we will have a reaction that may or may not be good. Should we treat a vine (Dionysian process) with an Apollinian plant (cypress, laurel, etc.)?

I have two spots at home which have received neither sulfur nor cooper for three years — teas have been sufficient. If I were to stop the teas, I might see diseases coming back on upper leaves. As noted previously, very little is needed: 300 grams of garlic and salvia by hectare is more than enough — perhaps too much! Interesting books by Maria Thun on the use of teas are also available.

Of course cryptogamic diseases can have diverse and varied causes, and are not necessarily linked to the phenomenon that we are describing here. Horsetail should not be thought of as a universal panacea, but rather as a supplement that restores balance when nature gives expression to one of these specific qualities too forcefully. In the majority of cases, these diseases are the consequences of our agricultural practices. In all, simplicity.

Anti-Acaridan Treatments Also Destroy Predators

The red-and-yellow spiders remain a problem for French winegrower. They can weaken the vine to the point of diminishing the alcoholic content of the vintage by one-half to one degree. Three anti-acaridan treatments per year are sometimes necessary. At this point, we must salute the marketing performance. The acaridan family contains about two hundred types, only two of which are dangerous for the vine. Among the 198 remaining types, we find the typhlodrome, which is nothing else but the predator on the other two, on which it feeds daily.

Like many useful beings, the "typhlo" is rather sensitive. It was not able to withstand the DDT treatments used in the 1950s. The rebellious spiders profited from this absence to occupy the terrain massively. Ever since the red-and-yellow spiders have been feared by the winegrower and the agrichemical industry found ways to exploit this new market.

Twenty years ago at the Coulée de Serrant, with the help of a representative of the Department of Protection of Vegetation who was sensitive to the balance of nature, we replanted typhlo eggs on four rows of vines. From the second year on, balance was nearly restored throughout the vineyard. Counting by parcels of land showed a strong increase in their number over a five-year period. The vine-stock cut in winter, with a little bark from wood — at least two years old — in which eggs are laid, allowed us to replant them on 7,500 acres. All that is required is to attach a piece of vine shoot at least four inches long to a vine every five yards, every two or three rows. For fifteen years there have been no anti-acaridan treatments on our domain. But remember that typhlos are sensitive and can be destroyed by doses of sulfur exceeding 10 kilos per hectare. They tolerate treatments with liquid sulfur better.

Can We Get Rid of a Parasite Without Understanding the Reason for its Presence?

The grapeworm is more difficult to deal with. This butterfly of dusk is a danger to grapes when it is in the larval stage. If one sides with the restrictive conception of biology, the problem is not insurmountable, since the bactospeine bacteria makes its eggs abort. And the new treatments, used in the form of "delayed action," allow the hatching period for the eggs to be covered. It is nevertheless profoundly antibiological to inundate one's vineyard several times a year with bacteria by the millions. How does that affect the laws of balance? Rudolf Steiner discussed these questions in his *Agriculture*, explaining the method used for the incinerations.

Incinerating an insect or the skin of a vertebrate on a precise date makes that type of life form flee at the end of the third year. In reality, they can no longer reproduce in this location. This method has an effect of repulsion. In the past, guardians of estates used to make a tar from the bones of harmful animals to keep them away, apparently with some success. Dates of incineration should respect a "planetary" calendar, since it is above all the reproductive forces of the animal that are targeted. And the animal, as we all know, is submitted to the rhythm of the seasons as far as reproduction is concerned. If one does not pay attention to these dates, the effect is nonexistent or very small.

In contrast to chemical treatments, which have an effect that is immediate but temporary, incinerations have no immediate effect, but have a residual effect that is infinitely greater. The way of proceeding with these incinerations is very precise: twenty or thirty insects, which can be dead, are put into a wood fire. When the fire burns out, the ashes are saved and dynamized for an hour to obtain a uniform mixture. The product can be spread in either pure or diluted form (usually a D8) on a basis of 5 to 6 gallons per acre. The date of incineration is the key to effectiveness and underlines even further the importance of backgrounds that generate life. One may also incinerate the seeds of a plant to prevent it from reproducing. For seeds this can be done whenever there is no date requirement. To have the fullest effect, it should be done for four years in a row, and it may even be repeated twice in one year. Very small quantities of these ashes are needed: $1/3$ teaspoon on the soil every 100 meters is more than enough. After applying, walk on the ashes to make sure they are not carried away by the wind. You will see almost no effect in years one and two, but in year three, you will begin to suspect that something is happening. In year four, the effects are obvious — the insects or seeds you have burned know they are not welcome in that place. If the effectiveness of these incinerations is incontestable, it is appropriate to scrupulously respect the procedure. To avoid the possibility that an insect (even dead) might escape in the smoke, it is suggested that it be done in a stove that has its door closed. The choice of firewood also may play a part, because it also has a planetary influence. We are in the habit of quantitative thinking, but in this case quantity is not all that important. Here we are focusing on a more subtle realm. It is this quality which, for example, permits a male butterfly to find a female at a distance of several dozen miles.

Before deciding to proceed with the incineration of grapeworms, it is useful to try to understand this butterfly, because there are locations where it never goes. A butterfly is deeply linked to the worlds of light and heat. In the dew of dawn, it is paralyzed. It must surround itself with a cocoon of light, of which silk is a manifestation, to leave the earthbound state of larva and gain access to the world of light and colors. But in this specific example, we are dealing with a butterfly whose peak activity is at dusk and we should take this into consideration in selecting dates of incineration. One can incinerate either the butterfly or the worm realizing that these are two totally different stages of evolution.

Here again, the date of incineration will change because we are not invoking the same creative forces, which is understandable from the point of view of earth/sun polarity.

When you collect butterflies with hormone traps (whose effectiveness will decrease in the course of time, as nature has a response to everything, so long as it is still alive), take care to place them far from your home, because they are not totally effective. Some butterflies will not get stuck in the layer of glue, and will go off to reproduce in your vines.

It is necessary to preserve some specimens for the following year, as the date of the first incineration does not always coincide with the first flights of the butterflies. The object is not to create an absolute vacuum, but to regulate and prevent invasions. The materialist conception of biology would be to have no trace of sickness. The living conception is to have them all, but to restrain them to a very weak level. This method of incineration is considered shocking by some. Let us explain to them yet again that thirty insects or the skin of a vertebrate are all it takes to regulate a species for one year. These animals are not burned alive. The poisons at present being used are much more toxic and have many secondary effects. Thus in the Maine-et-Loire region, the equivalent of nearly $140,000 in subsidies are spent yearly to control the muskrat. These products (these poisons) are particularly toxic, affecting other animals who feed on the cadavers.

Resorting to incineration does nevertheless pose a problem. Can one get rid of a parasite without having understood the reason for its presence and before having tried to remedy the imbalances which attracted it to the terrain? Sometimes these causes are beyond us, but that is not always the case, far from it. The appearance of the parasite is often the consequence of our agricultural activities. It plays a role in regulation. Parasites can try to suppress an excess of sap or simply bring an animal presence to vines growing on soil which had been totally deprived of animals for a long time. In this case, the incineration opposes a "mission" whose importance should not be underestimated. Incineration is a two-edged sword. It should be used with awareness, for we know of no antidote. In the agrichemical vineyards, the first effects seem positive, but later on can generate more violent disorders. Sickness is usually the manifestation of an imbalance. The plant or the soil gives a warning in the form of a parasite. Recommending treatments (whether chemical or biological) that are directed at the symptoms of disease rather than the causes is equivalent to working on the image you see in a mirror. A delusion, a lie which leads the plant sooner or later to a more profound level of rebellion, more pernicious, with an expanding complexity that always escapes us. It is perhaps by seeking with this method that we will understand the origin of new diseases.

The French Decree of May 1994
Wiped Out Years of Agrobiological Research

The policies adopted by certain organizations in the fight against new diseases are surprising. One asks oneself if the government decisions have not been manipulated. Indeed the terms of the Protection of Vegetation Act, a ministerial decree in France dated May 1994, discreetly made it possible to treat all the vines in France with highly toxic chemical products without obtaining the consent of the landowners. No sooner was it said than done. In southern France 10, 20 or 30 years of agrobiological work were destroyed in a few minutes. Since then it is no longer possible to prove that another way exists than intensive use of agrichemical products. The evidence has been destroyed. Seventeen-hundred acres had been treated in this fashion at the time of the official announcement of the decree. The testimony of an agrobiologist who became a victim of this measure was published in the local press: "In a few minutes they had sprayed everything from helicopters, even the chickens and the rabbits. I just had time to call my dog inside." That is where we are, in a system of unilateral thinking based on the protection of private interests. This decree is deeply illegal as far as Europe is concerned. It is strange that this scandal mobilized only a few winegrowers. This illustrates the carelessness we live in the midst of. In 1995, this government decree permitted the treatment of several thousand hectares, this time in the Midi and Bordeaux regions. Hundreds of thousands of acres will be affected in the near future. And the vines are not the only plants in danger.

What type of reasoning justifies such decisions? In this specific example, it is the "golden-yellow" disease which is in question. This disease discreetly made its appearance among the vines toward the end of the 1950s. Thirty years later, we are confronted with a major epidemic. The official statement is simple in appearance, almost convincing. We understand nothing about this new disease. We know only that stinging insects, in particular the grasshopper, can contaminate vines that have not yet been affected. We immediately deploy our insecticide weaponry to wipe out these grasshoppers. Whether or not they are biological, these treatments are being imposed on all of us.

A new decree in 1996 focuses on the varron (warble fly), a fairly inoffensive little parasite on cattle skin. Once a year, every French farmer is required by law to give their cattle an oral insecticide whose dangerous effects are well known. The government representatives seem to act as salesmen for the agrichemical industry. It has the logic of a dream.

Forty years of experiments and failures have not been sufficient to make us understand that one insecticide always leads to another that is more powerful. It is impossible to eliminate an entire population of insects. Not only do the survivors multiply more rapidly, but they develop a resistance. To destroy them, it is necessary to resort to even stronger products, to use larger doses or more toxic combinations — products which wind up in one way or another on the dinner plates of the consumers. As for the decree of 1994, contenting oneself with treating only the vines is a rather lightweight solution. The grasshopper can turn up elsewhere and besides, it is not the only chewing insect. It would be useless to try to speak about the usefulness of the animal world for the vine with such mentalities. Today it is a subject about which jokes are made. Yet again this problem, like many others, is being dealt with backwards. From there to admitting that these measures have been adopted in good faith and without double-talk is a difficult step to take. It is the entire continuity of a chain of life that is being destroyed. A chain of life that extends hither and yon in the soil and the living creatures. The health of the vine has been endangered.

A change, of course, would require recognition of errors perpetuated for more than thirty years that would put everything into question. Faced with such obstinacy one can ask if, in the final analysis, the objective is not to prevent the success of a new development of proven performance in agriculture because it puts in jeopardy very profitable markets. No one worries about the impoverishment of the farmer and his increased servitude to artificial products and authorities. The greater the dependence, the more the possibilities of withdrawal are compromised. These are questions on which the profession is divided and for trouble-makers it is a gold mine. This assistance to the world of agriculture is organized; it doesn't just happen by chance.

The indirect debt generated by such a policy is considerable. The word is not too strong — pollution of the water, of the soil and of the food chain — either directly through residues or indirectly by blocking the emergence of true quality. The trace minerals that were naturally present in our food thirty years ago now must be bought at the pharmacy. Should not the Minister of Health and Social Security react to this?

These new diseases are a sign that we have passed the point of no return. The gravity of systemic disorders is such that it has become very difficult to restore balance through healthy methods, as if in the final analysis that was the goal to attain. In certain cases, the ferocity with which government representatives imposed chemical solutions was flabbergasting. Thus we can present the example of an estate recently converted to biodynamics, where significant results had been obtained against the "golden-yellow" disease. The government representatives, although witnesses to the decrease of the disease, brought strong pressure to bear to have everything uprooted, which would have reduced to nothing the work of reconversion to biological agriculture. The reason given for this decision? The vines still contained mycoplasts, a type of virus that seems to be the vector of these diseases, so neighboring areas were at risk of being contaminated. Which is responsible for the disease, the virus or the terrain in which it developed? The well-known French scientist Claude Bernard said: "The virus is nothing, the terrain is everything." This hypothesis is today increasingly recognized by the medical profession. Agriculture prefers to ignore it. Who should be punished? The winegrower whose land is no longer vulnerable to disease because it has been brought back to life or, on the other hand, the one who failed to restore health to his vineyard, thus leaving it at the mercy of all sorts of imbalances, the one who refused an in-depth investigation and resists all attempts to change, even if the changes are possible? However the one who is the most to blame is the "counselor." Should not a government organism opt for durable solutions? Short-term reasoning always has the last word.

Are these diseases really new or have they been germinating for thousands of years? Every living organism is a bearer of patho-genic elements. The forces of health prevent their proliferation. They act in the end as a stimulant to the vegetative immune sys-tem. Should we reinforce the health, or try in vain to destroy all the microbes and viruses on the planet under the pretext that they are harmful? Should we place our vineyards inside sterile domes to protect them? The diseases attack the weakest plants because they no longer conform to their species and are inca-pable of generating quality. Sooner or later nature will take its revenge. If winegrowers are stubborn in their refusal of change, they are condemned to such an intensification of agrichemical use that the very concept of the "label of origin" will have lost its meaning. The "high" chemistry of the cellar becomes our only escape hatch, because the living system that generated the per-sonality of our wines has been definitely destroyed. The profes-sion will divide into two camps as partisans who defend the expression of life may be accused of being responsible for all sorts of contaminations. But let us repeat that it is the *terrain* which is responsible for the disease.

Is Eutypa a Symptom of Deficiency in the Circulatory System of the Vine?

Eutypa is a new disease which is more widespread than "gold-en-yellow" disease. A vine can even die of it in a few days. To put it simply, the canals of sap become blocked by fungus, but to limit the diagnosis to this is what a materialist would do. Is it not rather a deficiency in the forces of the vine which provokes stagnation of the sap? This sick sap becomes harmful and attracts the fungus which acts like an undertaker and does its work of decomposi-tion. The circulation of the sap is linked with the solar forces. How can a vine that has been saturated with systemics, nitrates and other insecticides, be as receptive as it was before? Under such conditions, it is always possible to spend astronomical amounts of money on research. Without a doubt, we will find a way to deal with this fungus. The problem will be momentarily resolved, until the arrival of another fungus, another plague, something even more pernicious.

Let us hope that the moment has come to deal with disease from a new state of mind. Let us select an image from another field — an artisan trying to restore a work of art, who limits his study to the damage to be repaired, is doomed to failure. A restoration worthy of the original work observes the entirety from a global point of view, before making the slightest gesture.

Portable Grafts or the Invasion of a Species

The phylloxera epidemics resulted in the grafting of all vines. This portable graft, rustic and thick, diminishes the quality of the wine. Let it grow, you will see how its leaf is different and its fruit without nobility. It is reduced to a tiny grain without a berry Should we congratulate ourselves on having with such grafting reduced the potential of our vines? Surely there was some other option open to us. Such a massive invasion of a species has one or several causes. In the United States the portable graft has not resisted any better: more than 150,000 acres of grafted vines were wiped out by phylloxera during the last few years. Why have they suddenly become vulnerable to the disease? Is it fate, or the consequences of a more worldwide weakening, brought about by all those years of false progress?

We have created a desert. Claude Bourguignon doesn't hesitate to write that in France the bacteriological activity in certain vineyards is weaker than in the middle of the Sahara. If you empty 95 percent of the apartments in a building, they will soon be occupied by squatters. Exactly the same happens after a disinfection of the soil: pathogenic elements are the first to return to the location. Only the life of the soil in the full meaning of the word permits each species to limit itself and live beside the others. Rudolf Steiner quickly confronted the problem of phylloxera, which he seems to analyze as a consequence of an imbalance between the influences of inferior planets (closer to the sun) and superior planets (farther from the sun). In order to understand the extent of this profound observation, take a look at the parts of

Agriculture devoted to the respective roles of these two types of influence, one oriented more toward faculties of vegetative reproduction, the other toward its quality (color, aroma, etc.). This requires a long period of reflection. But during the last century, when everything was organic, how did it happen? The source of real agricultural knowledge has been silenced since the end of the Middle Ages, an era when it was known that one man could not cultivate more than five hectares (about 12 acres) of land. It was on a surface of this size that he expressed himself the best. This figure is not neutral, because we find it again in the proportions of the sculptures of Leonardo da Vinci. In the last century the vineyards were in a state of exhaustion. It was not then admitted that we should "grease" the vine with manure. From a qualitative point of view, the results obtained were exemplary, but as soon as the vine's threshold of resistance has been passed, everything brutally collapses.

For centuries winegrowers always reproduced the plants by layering. A vine cutting is put into the earth to take root and become a vine. These vines were thus in reality hundreds of years old. They should have been regenerated by sowing seeds at precise dates, after having picked the grapes at the right moment of advanced maturity. The force that vegetation imprints on its seed, the vine puts beyond its seed, into its flesh, which explains those overflowing fermentations. The seed does not germinate well naturally. This difficulty favored the general use of layering. What type of wine would the vine of today, born from a seed, give us?

What is a clone but monoculture pushed to its paroxysm? This purely technocratic conception, which leads to the reproduction of a single specimen — supposedly the best — in dozens of millions of copies, does this not remind us of Hitler's ideology? The original definition of "best" was based on criteria that were strictly quantitative, and these criteria remain an important factor. It was a question of doubling or tripling the harvests. On the practical level, the solution could appear satisfactory: all the vines identical and isolated from forces that might differentiate them, are perfectly synchronized. Thus it is possible to harvest with machines and have uniform maturity. But who chooses the profession of wine growing because it is easy? A vineyard is like a

View of the estate taken from the top of Coulée de Serrant vineyard.

One of the many old walls separating the fields of wines. At right, plowing by horse on the grand clos, or biggest field of the Coulée de Serrant. The narrow vine spacing of these older grapes and steep terrain make use of horse power more practical than using track units. These grapes date back to 1920.

This former monastery, above, was built in the beginning of the 12th century by the Cistercian monks who first planted Coulée de Serrant. The 18th-century Chateau de la Roche aux Moines, below, was built on the ruins of the castle destroyed during religious battles about 1630. At right, heat emanates from a maturing compost pile. Cow dung, horse dung, and bio-dynamic preparations form the basis for the vineyard's composting.

At left, working with a horse in a field above the Loire. For the narrow width of the rows — these are the oldest vines on the estate — horses remain quite practical. At top, a field being plowed in the fall. Below, replanting vines in the spring.

At left, each vine is planted by hand. In the rough, rocky terrain of this Loire vineyard, it is not otherwise possible. After planting, the holes are quickly refilled to better retain ground moisture and to protect tender roots.

At the time of grape gathering, the fog — which lasts until 9 or 10 a.m. — often creates the ideal conditions for botrytis, a fungal condition desired to improve the texture of white wine. Two hours later the fog has disappeared. At right, the grapes often are carried on workers' backs, as some areas of the vineyard cannot be reached by track units. This practice also minimizes soil compaction.

An old well on a field facing the Loire. At right, the author is often accompanied in the vine-yard by faithful companions Flamme, or Flame, and Lune, or Moon.

To best enjoy a bottle of the prized Coulée de Serrant, the author recommends decanting of the wine several hours before drinking. This wine has been called the finest wine of the Loire Valley and one of France's best.

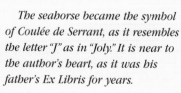

The seahorse became the symbol of Coulée de Serrant, as it resembles the letter "J" as in "Joly." It is near to the author's heart, as it was his father's Ex Libris for years.

Wine tasting in the drawing room. The author is pictured with his mother, Denise Joly. Coulée de Serrant was saved from the degradation suffered by many once-renowned vineyards by avoiding the mistakes of soil-destroying chemical agriculture. Behind is a tapestry dating to the 17th century depicting a hunting scene.

This well in the courtyard of the main house dates back to the 17th century. The age of the facilities at Coulée de Serrant underscores the awesome responsibility of maintaining its reputation for quality.

classroom, in it can be found a lazy one, a sleepyhead, a hard worker, an intellectual, and a clown. All sorts of individuals who form a class of their own. And the entirety of these unique and different classes creates the spirit of a school.

Isn't it the same with wine? Each wine has its qualities, often somewhat concealed, which can be complementary — resistance to disease, precocity, and shape of the leaves. A true selection would retain a few hundred samples out of several thousand specimens. One would first have more of a chance of bringing together in the selection subtle qualities not always immediately noticeable. Today they have the nerve to speak of such a selection (*selection "massale"*) when bringing together a few wretched clones.

These clones have also led to the practice called "green cropping," which means cutting the grape in July or August when one sees that the harvest will be too large. This way, one is sure to have the maximum desired yield, and the "excess" is destroyed. This is a terribly materialistic approach. If a vine have been "preparing" itself for a certain crop, destroying part of it few weeks before harvest must have some side effect. Nature can try to balance such problems, but at such a crucial time there is little it can do. A plant is not a machine. Now I hear people talk of "that green crop" to show that they use a "qualitative" approach!

The key factor in making a good wine is to understand nature and help it — one becomes nature's assistant rather than "a wine maker." Would you call yourself an "egg maker" if you raised chickens? We understand so little of all that is called "nature," of these incarnating forces. Can we truly claim that we "make" wine?

It is in this spirit that the selection of vines generating shoots is carried out. The oldest ones, adapted to the area sometimes for centuries, were destroyed and replaced by clones. However, they represented a patrimony which only time can reconstitute. The vines at the Coulée de Serrant, a vineyard for over eight centuries, were started over again with indigenous stock that dated back to 1920. These vines are very old and are worked with horses on a steep slope, so obviously not all of them make it. This astonishes our specialists habituated to a more mathematical conception of wine growing. One should not sacrifice the foundation for the form. This parcel of land is a precious nursery for vines and shoots.

When planting vines that are not clones, it will take five years to produce a first crop (two years longer than with clones), and you will have lower yields. People often ask winegrowers how many buds they leave on the vines when they prune it — first they should ask if the vines are from clones. With the same pruning, clones can produce twice as much — but they may be exhausted at the age of 20 or 30 years, while other vines may last for 50, 70 years, or more. It take often 20 years for the roots to reach their real nature (Dionysian).

A criticism could be made of vines grown from seeds because of the possible interferences at the moment of the fertilization of the flower. It may be fertilized by a different type of vine from the neighborhood. In a sufficiently large field, the risks are small and it would be possible afterward to eliminate these cross-bred vines. We could think of using these regenerated vines to replace portable grafts, as they would perhaps have a natural resistance to phylloxera. We would have, for example, above the soil a vine shoot issued from our older vines, grafted on regenerated wood grown from one of our seeds. The history of the location will marry with new blood. Isn't it time that research budgets consecrated to wine growing were made use of intelligently?

In 1994 we developed the planting of shoots in the vineyard. Growth has remained so slow 10 years later that the experiment cannot be considered a success. It should be tried elsewhere, however — the Coulée de Serrant has been planted with wines for almost 900 years in succession, so the soil is tired. Another reason for our lack of success in this endeavor is that the link of the vines to their archetypal forces is weaker than it once was.

It is difficult to foresee the taste of a wine made from a seed-grown vine, even if such an experiment can be achieved. At least we know that ungrafted vines — we've had a half-hectare here for 10 years — make a better wine than the grafted plants, although it is interesting to note that they have lower yields. Our local birds have made their assessment — whenever they have the choice, they obviously prefer to eat grapes from an ungrafted vine before considering the others.

When you graft a vine, the marriage is never harmonious: the difference in diameter below and above the graft is the sign of a forced union. Imagine that starting today someone other than yourself decides what your daily meals consist of. This choice belongs only to your own feelings. The relation between light and heat assimilated by the aerial part of the vine and the earth part taken by its roots can only be confided to one single entity for a perfect balance.

The Violation of a Harmonious Whole

The considerations that led to a new drama, whose consequences will be even more serious, are from now on united. Our era is preparing to sanction the rape of a harmonious whole. It is not so much genetics in itself that is to blame here, but the nearly total lack of awareness and understanding with which the subject is treated in agriculture. This theft of genes from each type of vegetation and animal (oaks, chestnuts, fish), which are then imposed by force on other species without regard for their respective roles and harmonious complementarity of realms and species is a violation of the natural world. Each piece of the puzzle, whose entirety is no longer visible to us, is sliced up in a blind and arbitrary manner. The world's image is shamefully defiled. Do we still have a choice in the matter? Thirty years of false agricultural progress have led us into a blind alley. The forces of consciousness which permit us to perceive our environment from a worldwide point of view are destroyed from infancy through childhood. What is taught to students is based on theoretical knowledge that deliberately camouflages the essentials. Under such conditions genetics can be imposed on us. There is also a lot of money to be made from this. Sooner or later it will be presented to the world as the miracle solution.

And those who have understood that this is just one more lie, designed to complete the work of destruction already begun, are regarded as pitiful kooks. A living being created by artificial procedures, and thus in contradiction to the laws of nature, no longer belongs to the collective organism which surrounds it. To get the animals to tolerate these new products, we must trick their protective instinct, just as we have already tricked them by adding synthetic aromas to poisoned flour. The deceptions keep on increasing.

Although fine reports concerning the new transgenetic plants, which are resistant to herbicides, to viruses, to predators, to fungus (what a program) have been submitted to high officials, the botanist Jean-Marie Pelt called for a moratorium "to take a little time to study the eventual consequences of these genetic manipulations. I remember when I was a student that DDT was presented as the perfect insecticide. It took 30 years to make it illegal."

Professor Gilles Eric Seralini, one of the four experts for the European Parliament on genetics, also considers that this story is a huge dead end. The real problems and side effects generated by this technology are always put aside and economics prevails. How can you ignore the system that originally put these chains of genes in a specific order? Genes coordinate between two worlds: one physical, the other pure energy. They convert one form into another. If the gene order is disturbed, conversion will most likely be poor or incomplete.

Knowing everything about a Boeing aircraft (a microcosm) and nothing about the atmosphere (the macrocosm) can only lead to a crash. The same will happen to our health and environment if these terrible practices continue. The forces behind nature are much stronger than any economically oriented mind! If we do not open our mind to these realities, then the side effects of our actions will eventually *force* our understanding in a drastic way. Look at the huge increase of new diseases in children, for example.

We were told that a nuclear accident was impossible, and then we had Chernobyl. This time what they are attacking is ecology through greed for profit, to validate the already large investments that agribusiness and growers have made without taking precautionary measures.

Man plunders nature, appropriates a process, has it patented and often applies it in an area which is not appropriate for it. We are thus going to generate new species, giving the pig an udder like the cow so that the piglets will grow up more quickly, or manipulating the bees so that their honeycombs will be larger and improve the production of honey, or putting a pig's heart into a man. The list is a long one, consult it yourself. And for a large number of our fellow citizens, undoubtedly sheltered since birth from nature's wisdom in a somewhat sterile atmosphere, in the image of our clones, all this is quite convenient.

The belief that a reengineered vegetable is immune to disease is a farce, for the laws of harmony are ineluctable and are capable of imposing themselves through violence if they need to.

We are in the midst of paying to learn it. Of course, these vegetables will be of interest commercially, but the taxpayers should know they will cause drastic effects on their surroundings, for which they will be obliged to pay as has been the case in the past. If a cell isolates itself from its organism because it is no longer in resonance with the organism, no longer feels linked to it, no longer listens to it, or can no longer function in it, it can perhaps start proliferating without respect for its surroundings. Is that not the case also with cancer? Life is a whole in which the elements listen to each other. Give them the possibility of doing so. Would it not be simpler to listen to the whole entity, rather than attacking the tumors with rays which destroy all forms of life, both good and bad? Is it not necessary to understand life in order to preserve it well? Can't we ask these companies promising miracles to face the huge cost of the possible side effects of their products on health and planetary well-being?

Last year a new species of tomato which cannot rot was introduced in France. Many professionals were planting it to improve their profit margin. In the end, these crops turned out to be unsalable because, although their shelf-life was indeed vastly extended, the fruit tasted rotten! Yet no one seems to learn from this lesson. Instead, more investment and research will be put into finding out how to get rid of the unpleasant taste. There is no future for humanity if we cling to these attitudes, only illness and self-destruction.

A Deeper Imbalance

"There is one who knows everything and understands nothing, and another who knows nothing and understands everything." Charles de Gaulle was speaking at the time of two of the Cabinet Ministers in his government. Everyone will agree that in the first category there is an overabundance of candidates, while the emptiness of the second category is dramatic. And it is precisely in this category that we have some chance of meeting sensible people. Is it possible to get out of the terrible blind alley into which our agriculture has been driven, while the power remains in the hands of humans who behave like clones, who all came out of the same mold? Technocrats share the honor of a uniform diploma whose contents are based only on criteria of intelligence and profit. Who can seriously pretend to evaluate the knowledge of an individual from his responses to a multiple-question test?

Hitler was a peasant, a dentist, a dictator, or a swimming teacher. Check the correct choice to obtain your diploma, please. It is not the "knowledge" which is judged, but the degree of automatism of the future technocrat, his capacity for not thinking beyond permitted boundaries. The professors who actually try to introduce a different approach are immediately disciplined. Their promotions can be seriously endangered. The situation is also uncomfortable in the research centers for those who choose to work on the terrain and to leave their computer/simulator from time to time. A computer will never contain anything that was not put into it. It is useless to await miracles from most scientists of the new generation. Since the cradle they have been brought up in this state of mind. Nourished both symbolically and physically: the crystallization of powdered milk for babies compared to human milk gives us a small idea of what became of the vital forces that a young child should assimilate. And these new sanitary regulations imposed on artisans, to the point of driving them out of business, are a measure of the considerable degradation of the quality of our food products, at the mercy of the first pathogenic agent to come along. The conflict between bio and non-bio is completely surpassed. Nature also creates poisons. The awarding of a bio label, which was necessary in the beginning, is now somewhat ridiculous. Excessive potassium, nitrate from fresh manure, hybrid species, genetically saturated types of wheat can all be bio. Quality eludes the law because it belongs to the domain of the intangible.

Our cells are renewed every seven years. What contributes to this rebirth, if not what one eats, what one drinks, what one breathes, what one hears, what one sees?

Nothing can evolve in depth without creating a new way of teaching. To specialize an individual without having put him into resonance since childhood with his surroundings and thus with himself, without having underlined the permanent and more secret exchanges that pass by us, is to take the risk of making him a murderer. A specialist can be correct at the level of his mini-sphere and spread death without being aware of doing so. Because he does not know that his "mini" territory represents only a tiny part of an organism that is much more vast. Can we understand the human body by concentrating on one of its toes? The success of the Waldorf Curriculum as founded by Rudolf Steiner (see Appendix III), which presented a free teaching method based on the psychological evolution of the child, is the fruit of this budding understanding. An burgeoning infatuation with this method in Germany, Holland and the Nordic countries, is perhaps a reason to believe in the emergence of a new spirit.

Chapter VI
Homeopathy

Homeopathy is not biodynamics, but it is important to understand what it is and how it works. Homeopathy is a therapeutic procedure based on the teachings of the Greek physician Hippocrates, who was the initiator of clinical observation. His physiology is based entirely on the theory of humors (blood, lymph, yellow and black bile) from which the temperaments are derived. Their balance (or good proportions) constitute health while the excess or lack of one of them brings about sickness.

Similia similibus curentur means the treatment of like by like. The basic essentials of homeopathy were defined during the eighteenth and nineteenth centuries by a German physician, Samuel Hahnemann (1755-1843), from experiments and observations on the effects of quinine and other substances on the organism. Homeopathy consists of administering to the invalid extremely small dosages of a substance capable of producing in a healthy person symptoms similar to those from which the sick person is suffering. Artificially triggering a danger signal awakens the immune system of the invalid. This treatment obliges the organism to recognize its enemy and react to it.

Isopathy goes even further, since it treats evil with evil, while homeopathy cares for like with like. Thus against a burning sensation in the stomach, the sick person should ingest something able to provoke such a feeling, such as pepper, for example. In isopathy one uses a dilution of the substance which is directly at the origin of the burning sensation. Only a few experiments on mildew and oidium have been done in terms of homeopathy and isotherapy — this is something which is worth investigating further, and it is less expensive than conventional approaches.

Allopathy is a diametrically opposite approach. It consists of treating the sick person by provoking a state contrary to the sickness in order to destroy it. The organism is not stimulated, quite the contrary. The remedy substitutes for the body's natural defenses, at the risk of putting them to sleep, acting somewhat like a crutch. In the final analysis, chemical agriculture is based on the same principle and the long-term consequences are comparable. Also the prolonged use of chemical fertilizers discourages the roots of the plant from doing their work of assimilating elements from the soil.

Allopathy is a form of assistance that places the organism in a situation of dependence that increases little by little.

Homeopathy does not call on the physical properties of the product, but on its qualitative background. The substance used, which sometimes on the strictly physical level can be a poison, is diluted, in a procedure difficult to conceive of for our Cartesian mentalities.

What is a Dilution?

As an example of a dilution, take one gram of a substance and nine grams of water. This mixture is shaken one hundred times. After a pause of a few minutes, one has obtained a dilution, one called D1. To this D1 add 90 grams of water. On the same principle the mixture is shaken, an identical period of pause is respected, and one has obtained a D2. There is thus in this solution one percent of the original product.

To these 100 grams one adds 900 grams of water. One shakes and waits. One has a D3 with a one per thousand concentration. If the quantities become too large, it is possible to use 10 or 100 grams of the preceding solution while adding 90 or 900 grams of water and by this process one obtains a D4. In this example it is a question of decimal dilution, thus it is based on the relationship between one and nine; that is, one part of dilution for nine parts of water. At the level of a D6, the dilution contains a millionth of the substance at the beginning of the process.

Modern science, refusing the division of the atom, can no longer admit the infinite division of matter. We are told that life only exists through the matter that manifests it and cannot detach itself. Beyond Avogadro's number the active matter of a substance, whose dilution in water one increases, has disappeared. If one sticks to the strictly materialist viewpoint, the solution should not and cannot have any effect. However, very much to the contrary, the experiments have proved that the effects are amplified to the extent that one augments the dilution. Spray different lots of sprouting seeds with different dilutions of arsenic. Of course, in D1 (10 percent arsenic) or D2 (one percent), it is still matter expressing itself. But in D8 or in D30 (there are thus 30 zeros) it is another aspect of the substance that is expressed and affirms its existence, since physically there is no longer anything. However, the effects do not progress in linear fashion, but by cycles with phases of inhibition of the activity of the solution, which are not constant.

In Dr. Kolisko's valuable book *Agriculture of Tomorrow*, she gives the results of different dilution experiments (by lots of from one to 30), conducted according to a very strict protocol on animals and vegetation. In one experiment mice were fed different dilutions of silver nitrate. As time went by, the behavior of these mice changed completely. They always stayed in the shadow of their shelter without going out, light bothered them considerably and in the end they developed more pathogenic states associated with the eyes. Wheat grains soaked for a few minutes every morning in a dilution 7 (meaning a seven-zero figure) of silver nitrate produced a tumor of the throat after a few months. A dilution 23 (thus, 23 zeros) produced a shapeless animal. Similar results were obtained with plants: a dilution 8 of hydroxide of calcium produced seeds three times bigger than the controls.

The same experiment with copper sulfate generated tumors in every batch of mice treated with high dilutions. Many fascinating experiments have developed in this book, which accounts for a life's work. In her little book, *De l'homéopathie à la biodynamie*, assembled under the auspices of the Ecole Supérieure d'Ingénieurs et de Techniciens pour l'Agriculture, Lilian Boehrer studied, among other things, the homeopathic activity of sodium arsenate on the breathing of wheat germs (coleoptiles). The results obtained bring into relief the activity cycles of certain dilutions.

Thus on the D3, D4, D5 levels, the breathing of the germs is almost entirely inhibited, the D6 has no effect, while the D8 has a very strong inhibiting action. The effectiveness of homeopathy has been scientifically demonstrated. That is the justification for the sale of these products in pharmacies. However, some practitioners are disturbed by results which offend Cartesian logic and persist in denying the evidence.

Of course homeopathy is not a universal panacea able to cure all ills. It is an opening, for it requires a different approach. Its performance justifies a far more intense mobilization of research than it has yet received. But it is confronted with opposition to change because it contradicts the principles of a science founded on physical manifestations of tangible matter. No one wants to take the risk of bringing into question the system of thinking and the contents of scientific training. Finally, and this is not a minor obstacle, it is much less attractive financially.

The Use of Homeopathy in Agriculture

In allopathic medicine, dosages are simply increased to intensify the effect of a product. The agrichemical industry reasons in exactly the same way. In wine growing, if the effect of a product tends to extinguish (yes, nature has the ability to react), one simply resorts to stronger dosages or goes to an even higher stage by using a more toxic formula. It is impossible to deny the effectiveness of the method; it is unfortunately accompanied by an increased dependence leading, sooner or later, to a blind alley from which there is no escape. But can homeopathy be used in agriculture? It is important above all to avoid confusion. If biodynamics and homeopathy both focus on the same background, they are derived from different disciplines. Biodynamics is not to be confused with homeopathy, even if in certain situations it resorts to using homeopathy. Biodynamics uses decimal dilutions in a relationship of one to nine. Homeopathy can use these dilutions by hundredths, one dose of active material for 99 doses of water. It is addressed to a sphere that is probably further away, undoubtedly with stronger effects, that are perhaps less direct on a plant. It is appropriate in any case to remain vigilant, for one never receives an immediate and visible answer when acting on intangible levels. The world of vegetation is gifted with a great sensitivity on these subtle levels, which nourish it constantly. Homeopathy is thus going to play an interesting part. But it is a good idea before taking action, to ask oneself if the action is justified.

For example, can one try to check oidium or mildew by isotherapy on estates where imbalances have been and are still intense? In other words, can one "forbid" nature to play its regulatory role through sickness, before having tried to correct the errors which are within our reach? If it is difficult to control the effects of atmospheric or microwave pollution on vines that have been brought up badly, trickling with herbicides sprayed by a neighbor, we still have (though the latest government decisions give us reason to doubt it) authority over decisions made about our estate. It is contradictory to use a living and powerful background to forcibly keep a plant alive that one is weakening in other ways. On a property being reconverted, the answer is easier to give. Thus the parasites, whose presence is linked to an excess of nitrate (even if it has been corrected) can remain for some time on the terrain. In this situation, an incineration of parasites according to the rules of biodynamics and a homeopathic dilution of their ashes would be reasonable actions. Let us retain in our memories a principle which has never been denied: *sooner or later nature reacts to being mistreated.* Our society has polluted the physical world. Is it not dangerous to open up to it the doorway of a more subtle world? Fortunately the very limitations of our system constitute a relative protection. A D8 contains only a hundred millionth part of active matter. No one will make their fortune by commercializing the basic product. Also a dilution in water has a short life. This requires that each one make it for himself at the time it is required. There is not much to sell. The dilutions sold in pharmacies are made with alcohol, and remain effective longer. The choice of alcohol as a dilutant is not possible in agriculture, because the quantities required are too large.

There is sometimes confusion between homeopathy, which awakens a process of defense, and the simple dilution of a substance that one can carry out on a subtle level. This procedure derives in the final analysis from an allopathic approach, since it is a question of giving a plant what it lacks, silica for example. It is then perhaps preferable to bring this substance into a dynamized form, in which its effect would be intensified. Up to the present time, the use of homeopathy in agriculture is nearly virgin territory. It is a field in which research is nonexistent.

Putting Homeopathy into Practice

Lilian Boehrer proved that it was possible through homeopathy to help plants eliminate excess copper which was poisoning them. This experiment was presented at the Fourth French Scientific Congress in 1964.

Seeds of dwarf peas were planted and treated with copper at regular intervals in such a way that their growth and seed production were stunted (reduced by 30 to 50 percent). This diminished harvest was divided into two lots. A third lot was made up of healthy seed. They were put to sprout in a room maintained at 23 degrees centigrade. The procedure required 24 hours of soaking, three days of germination and 10 days of culture. The water used on one of the toxic lots was a 15 CH (Hahnemann Centesinale) of copper sulfate. On the other lot, normal water was used. At the end of the experiment, Lilian Boehrer noticed that the batch of contaminated grain which had received the dilution eliminated the copper and attained almost normal growth. On the chemical level, the 15 CH solution no longer contained any copper. It retained other properties, which permitted the stimulation of the dormant ability of the plant to excrete the excess copper.

The experiment can be tried on vines saturated with Bordeaux oatmeal; a saturation which caused a large number of winegrowers to adopt the systemic treatments whose pernicious effects we have already described.

Instead, we shall force the vine to recognize this excess of copper as a source of trouble. Think of it this way: if someone repeatedly disturbs you, you will react several times, but eventually you will begin to "screen out" this annoyance. Multiply the disturbance by 20, however, and you will certainly react again. This is what we

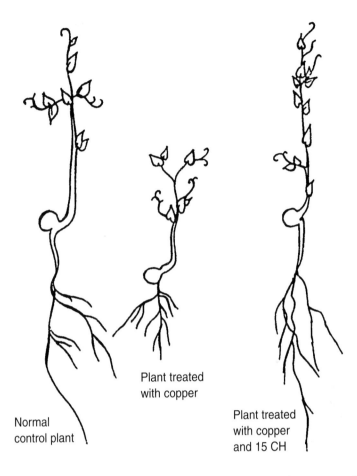

Plant treated
with copper

Normal
control plant

Plant treated
with copper
and 15 CH

will be doing with our vines. First, take 10 grams of copper sulfate, to which you add 90 grams of water. Shake the mixture one hundred times, then let it settle for three minutes. It is easier to use a syringe, such as one can buy in a pharmacy, for the following dilutions. One must retain the proportion of one to nine, but this time by volume. To the 10 cubic centimeters (cc) of the solution obtained, you add 90 cc of water, obtaining 100 cc of dilution. You take 10 cc from this which you mix again with 90 cc of water.

The quantity is increased when the moment comes to obtain the final volume desired. About 5 gallons to spread on each acre, and one treatment every six rows are sufficient. It is a question of bringing about quality, not quantity. If the spraying equipment can hold more, it is always possible to put more in it, but the effect will be the same. In making the necessary calculation to carry out the dilution, it is simpler to focus on the total quantity required and to go back to the first manipulation. The choice of the dilution is a delicate subject. The effects evolve in cycles, with stages of inhibition.

Rudolf Hauschka explains in *The Nature of Substances in Agriculture* that each substance has its own rhythm. We have already stated that life is only a sum of rhythms. From the simplest chemical reactions to the most complex life forms, it's all a form of music — an earthly one.

Dr. Kolisko's book contains a treasure trove of information. She gives the indications for a large number of substances, stating precisely their optimal and minimal aspects. These phenomena of inversion have not been satisfactorily explained to the present day. They are evidently linked to subtle backgrounds which dominate each substance, and perhaps to their planetary background.

Perhaps we can understand more clearly now that the life-bringing matrix of energetic forces or frequencies tends to die when locked up in matter. Matter is simply highly concentrated energy — just look what happens when it disintegrates. The more pronounced the "harshness" of matter, the more life forces die in it. This, finally, is the aging process for a human being, and it is what Goethe meant when he wrote, "Nature has invented death for creating birth."

The more one dilutes a substance in a rhythmic way, then, the more its force is removed from the specific matter in which it was caught (or dying, if you prefer), and the more it is restored to its intangible, archetypal, more active form. This means that the more we dilute a substance, the more powerful it becomes!

It is important to understand all this to reach a deeper understanding of biodynamics. Biodynamics is not just spreading a few products at specific times. It is mainly the task of achieving a deeper understanding of the natural system within which a farm or vineyard is located and from which it receives its life forces. Once this understanding has been achieved, then one can decide to balance or to reinforce one side or another. In some cases, for example, one would want to develop the link to earth forces (the physical side); in others, one will want to reinforce the cosmic realities (taste, shapes, etc.). One can also work on improving the autonomy of the ecosystem to achieve a balance that forbids the presence of diseases. This kind of approach would restore farming to its proper status as a form of art.

Example: D8 of Copper Sulfate

In D8, 300 liters of the final product are necessary. In D7, you need 30 liters of dilution to which you add 270 liters of water (always 1:9). In D6, three liters of dilution and 27 liters of water. In D6, 300 cc of dilution and 2.7 liters of water. In D4, 30 cc of dilution and 270 cc of water. In D3, 10 cc for 90 cc of water. From the hundred cc obtained in D3, one takes 30 cc for the D4. In D2, 10 cc and 90 cc of water. In D1, 10 grams of copper sulfate and 90 grams of water from which one takes 10 cc for a D2. To have the 270 liters (60 gallons) of water, it is easier to fill a 300-liter (67-gallon) spraying machine and to take out three 10-liter (9-quart) pails of the dilution. If precision is important, it is not so to the milligram here. One or two syringes are necessary, a one-liter receptacle, a five-liter receptacle, and a 10-liter pail. Let us remember that we do not dynamize the solution at each dilution, but we do shake it. Some practical details should be improved. For example, the pump of the spraying equipment, which turns the contents continuously in the same direction, does not conform completely to the scrupulous method of elaboration. It is nevertheless true that the effects persist.

It is impossible to discuss homeopathy even briefly without mentioning the Weleda and WALA laboratories. Weleda products are available in pharmacies throughout France. They provide a very good base of departure for dilutions whose basic ingredients you do not happen to have around the house. I think, for example, of a high dilution of thuja or calendula to cleanse an estate which is being converted from a past heavy with chemicals. The remarkably scrupulous procedures of these laboratories make no concession to economic demands.

Chapter VII
The Cellar

The domain of the cellar is not directly concerned with the bio-dynamic method, but is a prolongation of it. And it is only in this sense that we are going to discuss here the work that is done in the cellar. If we allow ourselves to penetrate the sensitivity that the vine has for its environment, if we have respected its capacity of manifesting all its subtleties in its sap and beyond that in its fruit, in short, if our agricultural activities have been of the right kind, the work done in the cellar is approached with serenity. One can keep track of these developments by measuring sugar levels of the sap from spring to autumn with a refractometer. Our confidence during the vigil of vinification measures the extent to which our understanding has progressed. From then on, the collaboration with an enologist (a scientist specialized in the study of wines), liberated from the obligation of acting like a stage magician, will be harmonious.

The Vine Makes the Wine

Is it really indispensable to use cosmetics, however noble they may be, to bring forth a wine? Does one need outside support to give expression to a label of origin? I already hear the voices raised in protest to screen off such indecent remarks.

This work is essential, some will say primordial. It is up to man to dress up the wine, to embellish it. Such comments, which I have heard many times, strengthen my own conviction. I plead in favor of an entirely different reality: true beauty manifests spontaneously in simplicity.

It is on the land, in the middle of his vines, that the wine-maker "makes" the wine. A healthy agriculture offers two major advantages: the resistance of the vine to unfavorable conditions in autumn and the quality of the vintage. A vine which is not habituated to feeding on nitrates as it takes in water resists heavy rain better. The structure of the grape's skin, strengthened by quartz and silica, permits it to pass through a phase of healthy and stable decay before the gray rot appears. In by far the majority of cases, if the surfaces are not immense, there will always be time to carry out the harvest without panic. And is it not in the "bad years" that one can judge the work of the winegrower? A vine's vitality reveals its true character when weather conditions are difficult. When too heavily made up, a face does not look good in the rain. Biodynamics, applied intelligently for several years, has no need to blush before any vintage wine, quite the contrary. We demonstrated this on numerous occasions during the wine tastings organized at the Coulée de Serrant with experts whose competence is universally acknowledged.

This implies, as we said before, that the vines are well located. This means, first, that the soil is poor. A vine receives from its archetypal forces an enormous strength to pierce the soil and to cope with the harshest conditions — keep in mind that the Greeks, who gave the name of a god to each force they were feeling, related the vine to Dionysus, the son of Persephone, the "queen of the underworld." It is the forces of gravity that are underlined here. The Egyptians had a similar view. When we read about Isis crying and looking for Osiris, her husband, who has been cut into pieces by the naughty Set, what does it mean? It shows the mystery of incarnation, in which an intangible energy system enters the earthly level and is concentrated by gravitational forces into the physical diversity of our world.

Understanding this helps us realize that all living components on earth are constantly communicating. Like the pieces of a puzzle, together they form an image, or a harmony, or a music. This what was called in ancient times the "music of the spheres." Returning to the vine, then, we need a soil where the vine can struggle. If not, the vine will behave like a spoiled child, and the wine will lack some sort of tension. Too often, and for the sake of yields, vines are overfed, and there are always the ubiquitous sales-people offering cosmetics or technology to redesign wines. With small yields, however, the design is perfect and needs no surgery.

Finally, keep in mind that it is important to find a place where the vine flowers close to solstice, not too early. After the solstice, days become shorter, bringing about a process of contraction to produce seed. A vine is very strongly related to the four seasons of the year, and early flowering means that the grape will have to start contraction too quickly. In other words, the vine cannot achieve good contraction when the days are becoming longer, when the vine is being drawn by this spring process in the out-side world to build or create its physical body. This affects the quality of a wine and the length of its taste.

Once the vintage has been harvested, if our agricultural gestures have been appropriate, interventions in the cellar should be undertaken with respect and discretion. Our main concern is to assure ourselves that conditions remain favorable to the expres-sion of the qualitative forces whose participation we have invited during the course of the seasons. These forces are carried from the vine into the grapes for its seed, for its progeny. Let them express themselves in the cellar. By our intervention, we refuse the mani-festation of a harmonious whole, of a balance that should not be disturbed. Only short-term thinking would try to fake or imitate this evolution to rapidly obtain a wine that flatters the taste buds. Having confidence in the potential of his own wine imposes on us an obligation to let it express itself.

Our competitive work in the cellar is limited to the rest periods of the wine after fermentation, for example, to make a decanta-tion. We can watch beforehand the good breathing of what is being "hatched" at the end of fermentation, and we can quicken the mixture with a vital new rhythm by stirring it regularly. But, before anything else, let things take their course, as in a true democracy. If the teachings given during spring and summer have been in-depth and complete, the results will be promising. The expression "to make the wine" always seemed inaccurate to me. In reality doesn't wine have enough power, enough autonomy and

is not only an agent of transformation, but also and above all a link with an organism that developed it, thus with the vintage year and the AOC. It prolongs the reality of the year in which the grape was born. And only a native yeast can by right of birth play the role of relay. May all these originalities revive intensely and appear in our wine glasses.

The INAO had to reply to a demand by the United States, made in the name of GATT (the world organization of commerce). It was a matter of the right to reduce wines to powder form for commercial reasons. The idea is shocking. However, is there such a great difference between this procedure and a filtration by centrifugal forces or by the use of reverse osmosis (a procedure for extracting water from wine). The proportion of water in a living organism is not put at random. If you decrease the water content from a human body by only a few percent death might occur. The vine is of course not as sophisticated, but such an action may indeed disturb for example an aging process. So do you want this technology or not?

A door should be open or closed. The choice is up to us. France is rich in exceptional terrains, and if France respects agriculture, it doesn't need all these artifices.

The Fermentations

One of the most satisfactory results of biodynamics is the rapidity with which the fermentations start and their capacity to push to the limit the conversion of sugars into alcohol. It is considered that 1.7 kilos of sugar brings one degree of alcohol. With grapes grown biodynamically the conversion rate is definitely improved. This is normal because the life factors have been intensified. The residual sugar (sugar not converted into alcohol), which always flatters the palates of the neophytes, is almost totally avoided as fermentations are pushed further. It is encouraging to notice that one escapes from norms based on calculations of "averages" when the forces of life have not been stifled. Our agricultural activities have enriched an omnipresent system of exchanges in all reactions linked to fermentations. And this phenomenon goes far beyond the work with yeasts. The birth of wine itself is intensified and thus its character, if not its longevity.

In this context, for the white wines, the skin maceration so much in vogue during the last few years, become a useless artifice. Once again, they are made necessary by the deficiencies of agriculture. This prolonged maceration forces the extraction of colors and aromas, which should have been liberated naturally while being pressed. It flatters the wine, whose must does not yield itself, as it lacks concentration, maturity and balance.

But reality catches up with us sooner or later. These subterfuges cannot correct in-depth a negative way of life. An informed wine taster will always detect a heaviness or lack of nobility in the wine. To force an exchange that is not made freely generates a debt. Unfortunately in the middle term, these taste profiles are in danger of becoming the reference for an appellation. That is why a few clusters of purists work for the recognition of a new approach. Other winegrowers will not delay in joining forces with this momentum, bringing along with them everyone in the wine network. Cellarmen, restaurant owners and staff, and wine waiters are concerned because, as a last resort, it is a question of answering the demand of a consumer in quest of authentic products, which are sincere and thus inimitable. The color of a wine is very sensitively reinforced by biodynamics. Simple pressure is sufficient to manifest it. At the Coulée de Serrant, enzymes and other extractors are regarded, without ill-will, as obsolete tools, like the vestiges of an immense lack of understanding.

Fermentation is the counterpart of what happens at the moment of photosynthesis in our vines, which is nothing else but the transformation of light and heat into matter. In the cellar, we witness the opposite phenomenon with the liberation of heat and gas. There is in this something like a birth and a death. In this manifestation of life, the fermentation can also be compared to a strong fever, a rise in the temperature of an organism, which tries to restore a balance that has been momentarily disturbed. To suppress a fever by an artifice penalizes the forces of healing. To impose a constant temperature on our fermentations puts a restraint on the expression of life. If one is obliged to regulate the temperatures, then one must be on the watch for an evolutionary curve, as life never expresses itself in linear fashion. A 15-year-old adolescent who is 5 feet tall will not be 25 feet tall at the age of

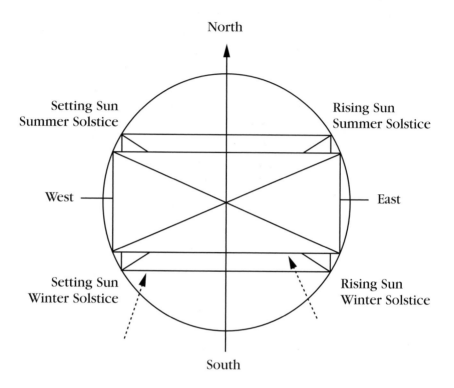

The four angles of these shapes are linked to the rising and setting of the sun at a specific latitude.

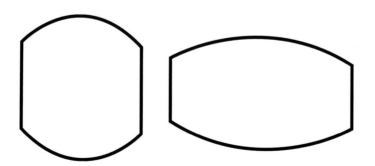

Traditional barrel shapes: the further south one travels (in the Northern Hemisphere) the longer the barrel's shape.

75. Nevertheless, some scientists reason in this manner when they decide so easily to date the volcanoes from 45.5 to 53.6 millions of years ago using carbon 14. That is to say, the carbon 14 measurement which, although correct today, cannot be applied to vestiges of the past whose reality was undoubtedly quite different. Was not the earth in its youth? Haven't its reactions evolved after all these thousands of years? Has the force of gravity always been the same? A static conception of things is purely intellectual. Let's not make the same mistake in our cellars.

The changes of temperature have become a necessity, taking into account the size of the vats used by the big vineyards. In the old days, the shape and volume of contents was adapted to each product, according to its profound nature, its vital forces. Look, there was a particular shape for a milk pot, for a water pot, for a wine pot. And each region gave them an originality. The relations of horizontality and verticality vary according to the shapes of the receptacles. This choice may not have been a random one. In the case of a cask, a double cask or even a 135-gallon hogshead, a brief rise in temperature to 80 degrees Fahrenheit never hurt the wine. With other materials in other shapes and quantities, the results could be different.

In the same way an interesting study has shown that up to the 18th century, most shapes of houses in France were linked to the points where the sun rose and set at summer and winter solstices. The more you move toward the equator (where days are always equal to nights), the wider the rectangle becomes — ending up with a line — the more you move north, the more you move from a square to an upright rectangle. This could help us to understand that barrels are wider and shorter in Portugal than in the middle of France. Shapes were a way to be perfectly tuned with a "local" reality. This knowledge existed for centuries and is being rediscovered and applied in agriculture to reduce labor, artificial inputs and costs.

The Dates of the Vintage

Between the planetary world and the world of vegetation, sympathies and antipathies are woven. Certain influences carry the upward thrust of a plant, others contradict it by opposing its aspirations. At harvest time the planetary situation which is dominant

Pressing grapes under the sign of the Balance *(ca. 1470)*

in the grape remains so, even when the grape is cut away from the vine. The living organism which was receiving the subtle forces (the vine) is not there anymore to transmit changes, so although the grape will be subject to changes, it will be much less so than when it was hanging on the vine. Thus, the date chosen for cutting grapes is very important. A harmony between the two "worlds" (earth and cosmos) is thus to be desired at this precise moment. This may be why people harvested when the sun was in the sign of balance, when gravitation and levitation forces are balanced. The fermentations will then be the exact reflection. The vine, whose finality is a fruit, is linked to the forces of heat. We will explain this more fully later, but basically it means a day on which the planetary situation of the earth brings more of these heat or fruiting forces. Thus, the fruit will keep more of its quality, and its development will improve. The dates of the vintage should be, as much as possible, more particularly marked with this tendency.

The advice is difficult to follow, because it requires stopping for a moment a group of grape harvesters. One could however find a middle term, by directing this team to an area that is less "prestigious" at less favorable moments. It will be an opportunity to watch differences of behavior in the cellar. One can also not take it into account. It is not a crime. There do exist a few rare periods when biodynamic agriculture strongly advises against harvesting. I think, for example, of lunar and planetary nodes (see Chapter VIII).

In Germany, Maria Thun directed a large number of experiments, the results of which found solid applications in the planting calendars. The example of onions is particularly interesting. These experiments are always conducted on a minimum of thirty specimens from each batch that is being tested. Depending on whether they are picked on a "root" day, thus an "earth" date, or a few days later on a "fruit" day, thus on a "heat" date, the duration of the conservation of the onions varied by more than two months. Of course the transformation removes the wine from raw vegetable matter, but other experiments result in similar conclusions, as for example on the rising of bread.

The work of Maria Thun has the merit of demonstrating in the context of a somewhat scientific protocol, that it is possible to amplify or, on the contrary, to restrain the vegetative resonance in which life expresses itself to the full. At the beginning and before going further, let's have a look at the phases of the moon during harvest time. During a rising moon or just before the full moon (two situations that can be different), the tendency is toward the formation of matter. It is an externalization. It is by the way sug-

The spiral illustrates how energies (non-matter) are concentrated into a point (matter).

gested that planting seeds be done during this phase because the germ comes forth from the seed more quickly. In the cellar the fermentations are tormented and more violent. The fruit-flies will appear to feed on this excess of energy, maybe in an attempt to regulate it. There are forces which are somewhat similar that we find again as spring comes. On the contrary, a waning moon is associated rather with the image of autumn. The fermentations are much more calm.

The influence of the moon is known to the great majority of the winegrowers in "living" agriculture. There are sometimes exceptions to these two rules, frequently linked with a specific planetary background.

Barrels for their Form, and to Hell with New Wood!

It is not my intention to stir up the traditional arguments about new wood. The subject has been debated many times. On the other hand, who takes an interest in the form of a barrel? This aspect is too often neglected. It is not by chance that in many cases nature bestows life in the form of an egg. A barrel can be thought of as an egg whose two ends have been sliced off, a vault when seen from any angle. Its form calls out to the peripheral forces that it concentrates in its surroundings. Architecture is the best example of it. Haven't you ever felt peaceful when in a room whose ceiling was half of a sphere or, better yet, a circular tower whose roof is a half-sphere? There is no comparison with the weight of the little cubes that have invaded our countryside. Animals that have been brought up with respect for their species and whose instincts have retained their integrity do not make mistakes. Put an empty barrel next to a dog's bed in a doghouse. Within a matter of hours the dog will have established permanent residence in the barrel. According to tradition, the Greek philosopher Diogenes, known for his disdain of honors, wealth and all social conveniences, lived in a barrel. The barrel is above all a call to peripheral forces and we deprive wine of them by putting it in a tank. This reflection on the shape of a container inspired the "inventor" of the wine pyramid which made its appearance during the Vinexpo a few years ago. However, it was not a question of accelerated aging, as was pretended, but rather a discreet harmonization. Still one must think of the polarities of the materials used and the orientation of one side of the pyramid to the north magnetic pole. This illustrates yet again that far-distant civilizations were highly perceptive of these subtle energies. They are often inspired by this living environment to establish precise rules in the realm of art. These traditions persisted for many centuries before dying out. For example, in certain societies music was considered sacred and only a certain caste could play music. It was perhaps understood as having come from a more subtle world that is endowed with organizing and creative facilities.

For the wine, the barrel is really the prolongation of the vine in the earth/sun polarities. As previously discussed, it is a shape which draws in or attracts a macrocosmic world. We can say that within this shape a macrocosm becomes a microcosm. What a precious shape! It is also the only container made of living matter. With wood, breathing is not cut off from the outside.

As for the new wood, everyone knows that its use has become commonplace. It is too frequently a factor in the uniformity of tastes. Each one makes his choice freely in full awareness, but there are perhaps limits that should not be crossed. Experiments are in progress with barrels made of exotic woods, which are naturally perfumed. Here again restraints have apparently become necessary. In biodynamics oak can be brought to the vine through a preparation, or decoction, of bark at the stage when the grape is forming. One can also carefully select a date for working and weeding, thus increasing the size of the pips and bringing the wine abundant natural tannin. Keeping an old barrel for 10-12 years is certainly possible by using hot steam for a perfect cleaning and to remove tartar, which stops respiration through the wood. This means that there is a maximum of 10 percent of new oak in the wine, and if the barrel is not overheated inside, that artificial taste of "caramel" is avoided.

A barrel of 220 liters is, in my opinion, too small, a bit like a 37 centiliter bottle. The best barrel size is 500 liters, which performs more like a magnum as compared to a small bottle. We should also mention amphorae, with their fabulous "Dionysian" shapes. The shape of an amphora is something like the shape of the steeple of an old church, which acts as an antenna for upward forces — except that the amphora is orientated downward, calling Dionysian forces. In ancient times amphorae were for the same reasons buried in the soil up to the top.

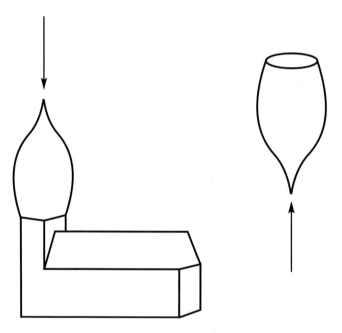

Cosmic forces drawn downward (left). Earth forces (Dionysian) drawn into the amphora.

Sulfur as a Process of Light and Heat

After the danger of "bio" becoming a fad, caring more about appearances than about the foundations and laws of life, the latest fashion is wine without sulfur. My first questions about such wines are: was sulfur replaced by ascorbic acid or by potassium sorbate? Or was the vine protected from refermentation by sterile filtration? I consider these three alternatives to be worse than sulfur. Is it possible to produce a good wine without sulfur, then, since the lees contain some sulfur? The answer is yes. But in such cases (and mainly for white wines), one should not send the wine too far away from the vineyard and only when the temperature is fairly low.

Should we follow such practices for the sake of avoiding few grams of sulfur? I do not think so. It is all a matter of quantities, and therefore of the balance that the farming practices have achieved. A few grams of sulfur applied two or three times is usually enough. A volcanic sulphur — which can be difficult to find — is certainly better than a manufactured one. Regarding synthetics: most people have forgotten that the process of fabrication is essential to the energetic aspect of a product — they think, for example, that a synthetic vitamin is the same as a natural one because they have the same "physical aspect." What a misunderstanding. One must look for the effects that this physical substance produces, not merely its basic chemistry. Different means of production or genesis affect the affinities of substances for energetic processes. This was one of the bases of alchemy: to move from one substance to another, one had to work with their affinities or their "memories," their ancient past, if you prefer. What is sulfur, if not a process of light and heat? It appears on volcanoes, there where matter is constituted. As for the vines, the powdered "flower" of sulphur is beneficial to them at their moment of flowering. How is it used in the cellar? In the old days sulfur was introduced by combustion, by lighting a wick. The effects of a combustion are far from being insignificant.

Rudolf Steiner spoke about how in agriculture the balance brought by sulfur through the yarrow (*Achillea millifolium*), which is, as we have seen, the base of a preparation. And it was not so long ago that winegrowers used to put a few yarrow flowers in each barrel as protection against oxidation. In his massive two-volume *Les méthodes biologiques appliquées à la vinification et à l'oenologie*, Max L'Eglise explains that the danger for wine coming from "live" agriculture is not oxidation, but reduction. This is certainly true, especially for biodynamic wines, where the racking has to be strong. Oxidation is an aging process; biodynamics is a youth process, so they are opposites. This goes along in the same sense as Vincent's electronics, in which oxidation-reduction remains an essential feature.

Is oxygen a friend or enemy to the wine? Let's look at an image: man needs oxygen to live, yet nevertheless this contact marks the beginning of the process of growing old. Thanks to his vital forces, he resists. His energy declines progressively after a full period of flourishing. In the same way a wine full of vital force will "flower" with the oxygen. But this too often only partly compensates for castrating agricultural procedures. One must understand that modern farming isolates a vine from its surroundings; modern synthetic molecules that actually enter the plant's sap (not simply remaining on the surface of the leaves, as was the case 20 years ago, before the systemic process) disturb exchanges between the leaves and the atmosphere, and also affect the quality of photosynthesis. Adding oxygen is very limited in balancing these disturbances. It has become standard to rely on technology to restore cellar oxygen in the wine (microbullage) — no one seems to understand that this problem is linked once more to improper farming practices. Winegrowers in biodynamics will laugh at the mention of adding oxygen to their wines.

In biodynamics, the danger is just the opposite: *i.e.,* reduction. Generally speaking, a biodynamic wine will benefit from a long decantation period, which underlines the vitality in these vines.

Aeration is an indispensable necessity for wine tasting. It is interesting to taste a little over a period of eight days, leaving it half-full in the corked bottle, without putting it in the icebox. We have often tried this at the Coulée de Serrant. Even if the temperature is 70 degrees Fahrenheit, usually wine passes through a peak period between the third and fifth days, before very slowly oxidating. And the quality of oxidation, its fineness, is also interesting to study. If you add a few grams of sulfur two or three times to the wine, especially when it is going to do some traveling, this is not at all shocking.

The Cold is a Force of Death

To let a wine get below freezing is a big mistake. Cold is a force of death. We may as well know it, even if this treatment is necessary for some wines. It is a question, so they say, of precipitating the tartars that are undesirable. If one really wants to get rid of this tartar, then the temperature should not fall below 4 or 5 C, and most will disappear. But one should also consider the role that tartar plays in the evolution of the wine, not just the taste at one point in time. Life is a whole.

Sticking Together

The procedure known as "collage" (sticking together) prepares a clarification. One could however feel a certain reticence about putting fish glue or the more noble method of the white of an egg in your wine, even if they are not intended to remain there. The egg and the fish are one thing, the wine is another. Can one then clarify wine without this "collage?" I reply "yes" without hesitation. These procedures should be above all considered as medications adapted to situations of imbalance, whose cause must be looked for further upstream. An excess of nitrates, a systematic treatment on the leaves, certainly has secondary effects. It is not unreasonable to think that they can result in naturally difficult clarifications. It is not my intention here to cast blame on a winegrower who is dealing with a situation for which he is not always responsible, but to provide a context for thoughts on a very much larger scale. And to understand once again that the further away we get from the laws of life, the more we will be submitted to artifices which originally were *not* indispensable.

Filtration

Is filtration really indispensable for our wines? Don't we also take the risk of leaving the appellation in the scum on the filter paper? The question may seem excessive, but it should be asked. It is possible to make unfiltered wines, which would certainly be better. At the Coulée de Serrant, when the residual sugar is low, we chose the solution of pre-filtering our dry wines on loose earth with a high silicon content. This earth brings a luminosity through its origins. It also has the merit of not robbing the wine. A sterilizing filtration, although it stabilizes, may have consequences on the aging of wine. Where would those marvelous '47s be if they had undergone such a treatment?

No removal of the lees, no cold treatment, no glue, no new wood, and only a prefiltration. What remains? The wine itself. It is up to each of us to judge if that is sufficient or not.

Wine Tasting

Before speaking of a "good" wine, shouldn't we take an interest in its truth and respect for its origins? Some rare, high-ranking tasters know how to evaluate these characteristics in their judgement. But for the others, tasting is a taboo topic. Nevertheless it is time for us to agree on a professional standard of ethics in this domain. The providing of samples is a question loaded with thorns. It is not a good idea for the grower to provide his own bottles. On some important occasions, the samples should be bought through commercial channels in order to be honest and fair. Why not even buy them in foreign countries to which they have been exported? Everyone knows that on some properties most of the harvest is done with machinery, and only several acres or less are harvested by hand. Which is the wine submitted for tasting, the first or the second? It is not a question of casting blame on the profession, but to support a silent majority of winegrowers who smile either philosophically or with annoyance at these masquerades. It is up to each of us to respect a code of good personal conduct on his domain under the condition of not being ashamed of one's own choices and under the condition of not disguising reality with cosmetics. Some tasters, ardently devoted to the vines, have the opportunity during their travels all over the world to compare wines. The differences are sometimes surprising. It is fortunate that the comments published in the guides usually present a synthesis of these different tastings.

It is also interesting to judge wine at different temperatures. If the cold can mask imbalances, at 70 degrees Fahrenheit all becomes transparent.

We can also, following the example of Jacques Puisais, taste wine under lights of different colors to realize that it influences our taste one way or another. With each color one has a deeper perception from a smaller part of the wine's spectrum of taste Only true light gives an authentic and complete perception. It is urgent to reflect on a revision of our rules for tasting. The public reputation of our AOC in France and in foreign countries depends on this rigorous attitude. And it seems these resources are already being squandered. Since the beginning of the 1990s, a page has been turned with authenticity being recognized as a necessary and major quality. Since we know some Asiatic countries now intend to require analyses to control the residues of heavy metals, herbicides and pesticides before allowing the wines to pass through customs, it is an indication that a change of course is necessary toward a sincere agriculture. France does not have a monopoly on the vine, but has in her hands advantages which are widely envied; a situation, a diversity of terrains, of vines, of know-how, of traditions. We are no longer able to afford squandering this capital without counting it.

The Cellar

How many cellars have been definitively amputated of their powers under the pretext of modernization? Reinforced concrete replaces the profound resonances of stones cut by masons who had the knowledge of polarities and placed the stones accordingly, in the direction they had before being quarried. Sealed into the walls, they found again the magnetic pathway of the vibration of this north/south magnetic current in which they had been bathed since the beginning of time. That was also the secret of a Stradivarius: the respect of the double polarity, both vertical and horizontal, of wood. Involving oneself in investments that are often excessive and ignorantly sacrificing a cellar built with wisdom is something to really think twice about.

In the old days architects and masons did not work blindly. Certain churches constitute evidence of this ancient knowledge, whose powerful effects we still feel. The builders of the Middle Ages chose a precise day to mark out on the earth the exact alignment of the sun in the sky, from its rising to its setting. It was according to the line traced and to other precise factors that they

constructed their edifice. Thus on a precise day each year, all the openings of the church receive an optimum of solar energy for they have been perfectly synchronized with this cycle, which occurs only once a year. Pilgrims flocked to charge themselves with these beneficial energies. The church was named after the saint whose feast day was on that date, a day favorable for the action of the solar forces of healing that the church was concentrating. If you doubt it, try to find an ancient church whose head is not situated between the north and east. And indeed, inside the Cathedral of Chartres which is unique among cathedrals, there is a labyrinth over which chairs have been deliberately fastened as if to prevent the labyrinth from accomplishing its mission. Today all this relates to what is called geobiology. That is another subject. Just the same, let's add that these energetic forces can be photographed by what is known as the Kirlian effect. In the rare churches that the 19th century did not devastate one can, by having someone pass through precise locations on a precise date, intensify his energetic forces whether or not he is a believer.

Certain cellars are treasures of wisdom through the designs of their vaults, their orientation, their proportions and the positions of their pillars. But pay attention because we are entering the realm of the intangible, the invisible, a domain that does not escape commercialism by smooth-talking hustlers.

Isn't it a shame to see reinforced concrete on the floors of cellars shutting out the exhalation which the sun at dawn brings to birth for the earth on that day and the inhalation offered by the setting sun in the evening? Wine is, without a doubt, sensitive to this rhythm. In a certain manner concrete prevents this breathing.

It is also difficult to deny the harmful effects of electric current; 50/60 hertz, or 50/60 vibrations per second, is a rhythm in profound contradiction with the laws of life since it is used, for example, to kill yeast in the industrial fabrication of cider. Recall the four states of matter discussed in Chapter II — it is important to recognize that electricity underlies these states. The physical level could not exist without electricity, or without a "plus" and a "minus" charge. It is through tiny doses of electricity that matter "coagulates" — but it is an earthly process, a material process, a mortal process, against which life forces have to fight.

In *The Nature of Substance,* Hauschka describes an experiment in which the flame of a candle is exposed to electricity, thus negating its upward movement. Electricity is a force deeply related to the earth, cutting away from the upward forces that bring life. According to Steiner, a child in its first year should not sleep in an electric field. If it does, the quality of its inner life will be certainly changed, producing a much more materialistic adult.

Detectors are available in the 50-60 hertz range. Most are not highly sensitive, and thus are useless, but a few are very good. Most houses are full of electromagnetic pollution. With a poorly grounded plug, a wire can have an electromagnetic field of several meters. When we are exposed during sleep it is worse, as we cannot defend ourselves. In a TGV (high-speed train) in France, each car is saturated with it. First it stimulates your nervous system, but then, after an hour or two, you sleep — just look at the number of people sleeping in a TGV. The same is true in many houses, which too frequently employ metallic structures. All those wires set into the walls inundate the rooms with harmful agitations for dozens of meters sometimes, even if the lights are turned off. Remember it was not so long ago that cellars were equipped with low-tension electric systems. Today it is possible to remedy this inconvenience to some extent by installing an interrupter on the electricity. This clever system of magnets cuts the electric charge in the wires as soon as the light is turned off. A direct line can be kept for the central heating, for example.

What is true for wine is also true for man. How many sleepless nights could be avoided if those phenomena were better known? Try the experience by disconnecting your electricity in the evening. Your night will be much more peaceful. Of course if you live in a building this will not work as nearby installations reach you through the floor and ceiling. It is a real problem which has effects on public health that are much more important than one imagines. And sensitive persons, those the earth has need of, are of course the most strongly affected by these phenomena. There is actually a very large lawsuit in the United States over this. Many cases of leukemia appeared after a high-tension wire was placed above a school. Although these cases have not yet been won, there is hope. In France, one high antenna for mobile phones had to me moved away from nearby villages, although their harmful effects were not officially admitted. The biggest court cases we will see in the future will probably involve the health impacts of mobile phones. I am told that AXA, the big U.K. insurance company, has removed from their contracts all problems that could be linked to the use of a mobile phone. Don't forget that some frequencies have become significant weapons for the army. Gigahertz-range detectors are also available, but they are bigger and more expensive. In Germany you can buy a net of fine copper that is used as a bed canopy to protect the quality of your sleep and health.

Twenty years ago there was a study in Germany which demonstrated that deaths of children under two years of age were almost ten times the number that is normal among subjects situated at less than 300 yards from a high-tension line. We have not come to the end of the unpleasant surprises which technology has in store for us. Thus to measure the effect of a current, all you have to do is to listen to your radio when you are in your car and driving by these lines. Is not the human being constituted of waves and rhythms that he receives and emits? What are the interferences with the others? Do you know that a microwave oven working with an open door can kill a Great Dane in less than an hour? Shouldn't we be more aware of the deep laws of life before launching new technologies? Indeed it is not unusual to hear a winegrower tell you that for a reason he does not know, the fermentations take place more violently in certain areas of the cellar and with more tranquility in others. These manifestations are even more frequent when the fermentation is not carried out in barrels. It is also worth noting here that if metallic containers are used, they must be grounded. Today, all metals connect themselves to this pervasive frequency pollution with which we are filling the atmosphere. The same is true in your fields — don't use metallic poles, as they concentrate this pollution on your vines.

There is thus concrete material to explore, an opportunity to deepen our knowledge of what we have under us and its effects on our houses and our lives. That is another subject, which also leads to enrichment.

Chapter VIII
The Planets

"If you want to understand the point, observe the circumference." — *Goethe*

"Man is a citizen of the universe, and a hermit on the earth." — *Rudolf Steiner*

It is entirely possible to practice biodynamics and to evaluate its effects without taking into account the planets. An in-depth knowledge of the planets is nevertheless a valuable asset in wine growing. A few simple experiments are sufficient to convince us, for the differences manifested are very perceptible. Nicholas Culpeper, the famous English botanist (1616-54), gives a planetary and sometimes even a stellar background to a large number of the plants mentioned in his works. For example, borage (*Borago officinalis*) is an "herb of Jupiter," under the influence of Leo, and the "chélidoine" (*Cheledonium magus*) which means "gift from heaven" should be picked "when the sun is in Leo and the moon is in Aries." Now that the symbiosis between plants and the cosmic world has been explained, we can understand how each moment has a particular quality for a plant. When you take into account the movement of the earth, of the moon, of the planets and, in the background, stellar constellations, every day on earth has an originality which might never recur. When a plant is "asked" to heal a specific illness, it is essential to choose the moment which "intensifies" the qualities of the plant. However this subject also remains taboo. Attempting to discuss it among an honorable assembly of winegrowers often leads to being made fun of and dubious smiles. Where does this lack of understanding come from? How can cultivated people conceive of our earth as totally isolated from the rest of the cosmos?

The mystery worries people, but we are surrounded by questions without answers. The infinite diversity of the world of vegetation is an enigma — these seeds which take different forms and bring forth matter in ceaselessly different ways. There are also mysteries in the animal realm where an embryo surrounded by albumen engenders a vast and surprising wealth of fauna. Does this diversity come from the earth and only from the earth? This earth who offers only a desolate face during the months of winter. The role of the sun is today recognized and we are ready, or almost ready, to admit the influences of the moon. But what would life on earth be like if the other planets did not exist? Not a single researcher has the key to this big question. No, as a general rule it is ignored and concealed, although a truly scientific attitude would be to keep one's eyes open on the unknown.

When a tourist visited Moscow before the fall of Communism, the guide who accompanied him never missed an opportunity to discuss the terrible poverty and total absence of liberty among Europeans. It was often impossible to convince him that the opposite was true, as he clung to a received idea, being a prisoner of the conditioning to which he had been subjected since infancy. In the same way education, such as is administered today, results in the sclerosis of our faculties of perception. Man remains deaf and blind to the universe. The exclusively materialistic training given to scientists considerably restrains the extent of their thinking and their investigations. Only the material criteria are considered to be worthy of interest. It is a dangerous initiative to seek an answer by going beyond the limits of the tangible into a more subtle realm, as it can be interpreted as a betrayal of professional ethics. A few rare individuals try to break free from these mental shackles. Their competence is brought into question and their careers are really endangered. See for yourself; unfortunately, I am not exaggerating. Why is there such unilateral obstinacy if not to protect the foundations of our society from an inevitable re-evaluation? Can we hope for the slightest spark of light from researchers in the service of the binary system of the computer programs? Before a response is given, it passes through a series of tests whose unfolding is invariably based on yes or no. A third solution does not exist. This does not leave much room for the study of quality, which is unable to enter through such a narrow door. How can you describe a landscape or transmit an emotion in such a restrained setting? There is thus no way out for science except to be beside the point, by limiting its observations to the quantitative aspect of quality. In his book, *From Jundishapur to Silicon Valley*, Emerson suggests putting a trinary system to work, in which man would have a place.

Life belongs to another less tangible world that is just as real and which interacts with the material world to organize it — of course, on the condition that life is not stifled. And nature still has the wisdom to render sterile everything which does not integrate with its organism, such as hybrids, for example. This invisible creative world, because it is not understood, is neglected and assimilated to an automatism. Is this not proof of the great ignorance of the laws which govern us? This world animates us, it breathes life into us. It is what differentiates a man from a cadaver. Its creative energies are omnipresent; its activities are manifested in our daily lives. Look at the power structuring a violin and its bow, which at each note prints a different shape on the powdered elder placed on a metallic plate. Where does this force that imposes itself on matter come from? The higher we rise in the hierarchy of the realms of life (mineral, vegetable, animal and human), the more the degree of autonomy is important when faced with these influences. It is an autonomy which culminates in the human realm. If, for example, the cycle of the woman is 28 days — the same duration as a lunar cycle — her cycle is nonetheless free from the phases of the moon. But this relative independence does not permit us to live for too long a time far away from the point of origin.

All the experiments in speleology have proved how difficult it is for humans to adjust to isolation and darkness. Beyond a certain length of time, life "under the earth" becomes survival for the human being. The effects of jet lag when flying from one continent to another show to what point we are marked by the solar system.

From Sky to Earth

The origin of life should be sought in a vast organism, of which the earth, sun and moon (among others) are all parts. "If you want to understand a point, look at the circumference it comes from," said Goethe. Ask our scientists if they know of a living system whose different parts do not interact with each other? This materialistic conception of life, which began to grow at the end of the Middle Ages and came to full prominence during the eighteenth and 19th centuries, was undoubtedly necessary, but it is dangerous when it is not balanced by its counterpart. This is particularly true in the field of science. I think especially of the Newtonian (non-Euclidean) aspect of weight and matter, based on the study of a material quantitative world, the only one to be really affected by the laws of gravity. The reasoning is not false, but suffers from being incomplete. Why do we systematically neglect the opposing influence of formative cosmic forces? This may be for future generations the most comic, or perhaps the saddest, question concerning our so-called advanced society. We have gone so far into error that it will be very difficult to avoid the enormous consequences of our actions. Steiner compared the earth to an elephant in captivity — you may be able to mistreat him for years without any problems, but one day he will break his chains! I fear we are very close to that point.

In their book, *The Plant Between Sun and Earth*, Adams and Whicher develop some new hypotheses. Basically, if a material pole exists which allows life to manifest on the physical level perceptible to our senses, the laws of polarity postulate the existence of an opposite pole for which matter is in some way negative. Building on this principle, Adams and Whicher oppose to the terrestrial pole another pole of anti-matter. This hypothesis leads them to reflect on the role of counter-weight that the sun and planetary system have upon the earth. By geometric projection they are able to explain the organization and energy centers which elaborate the forms of plants, shaping their leaves, their flowers, their petals and the disposition of their seeds. The many photos published in this work offer new perspectives at a glance and underline the questions that are too easily excluded. Such a complex organization of forms and designs, are they really the result of random chance, or the work of a hidden gene that the power of electronic microscopes remains unable to define? Isn't it like looking inside a radio for the person whose voice you are hearing? Is it really reasonable? No one can understand a man by focusing their attention on only one of his hairs. It is time to leave the infinitely small in order to turn toward the macrocosm and toward the different components of the solar organism to try and discover their complementarity.

This analysis is actually being carried out by a group of scientists in Russia, whose members defend the thesis of a cosmos that is alive and interacting with humanity and earth. At last an approach ready to consider that the infinities said to be voids in the cosmos only constitute a quantitative void and a fully qualitative approach answers to other laws. What a breakthrough. For a Cartesian mentality, it is impossible to conceive of the influence of a planet, a tiny scrap of confetti millions of miles away, on the different realms of life that surround us. Instead of concentrating only on the visible aspect of the apparently isolated planets, our attention should be focused on the sphere in which it expresses itself, the limits of which are marked by its orbit. From this point of view, the solar system is an organization of circles or of concentric spheres, with multiple interactions at all frequency levels. If we were to try to summarize Rudolf Steiner's analysis of this subject, it would be an impossible task. However, he talks of the genesis of our actual earth, which began with a slow process of contractions of a body of calories, thus of heat. After very long periods of time, this body contracted into a smaller gaseous body. In a third step, the luminous, gaseous body contracted again into a fluid, watery body. Then the earth progressively attained its present state of hardness. At this stage, the water diffused in a still opaque atmosphere, condensed and separated itself from the air. This period could be assimilated to what many religions and mythologies refer to as the Flood. The oceans filled up. The elements truly differentiated themselves from each other. Life in physical form could now appear. To each stage of this genesis corresponds a state of matter and the dominant influence of a planet. Saturn and Mercury mark the influence of the forces of heat, Jupiter and Venus the influence of the forces of light, Mars and the moon the influences of the forces of water. This analysis has been presented too briefly, but casts light on the eventual origin of the four states of matter that are present everywhere in our environment, including the organization of the layers of the atmosphere. If one is willing to give some credit to this genesis according to Rudolf Steiner, it is easy to conceive of the influences of each planetary sphere on the different caloric, atmospheric and aquatic states which surround us. It is also easy to understand why the position of each planet in relation to the earth can act through the four states of matter specifically on the leaves (water), the flowers (light), and the fruits (heat). This approach, by the way, joins the hypothesis of a group of astrophysicists that in the far distant past Earth and Mercury were both a single body.

The gradual arrival of a state of hardness can lead to an entirely different conception of biology. The books by Walter Cloos (*The Living Earth*) and G. Washmut (*Evolution of the Earth*), although they are not very recent, would of interest to those seeking insight on this point. They conclude that life does not come from the inorganic, but on the contrary, it decomposes into the inorganic after it dies. To try and explain life by starting from a material base may be as hazardous as trying to reconstitute an artist from his paintings. The pyramids also remain a mystery for the world of science. How did the Egyptians achieve such a decrease of magnetic fields in the King's Room? How did the they cut those stones with such precision? The author of an article that appeared recently in a scientific review estimated that only a laser could achieve such a level of precision. If, as Rudolf Steiner says, the state of hardness of the earth and thus its gravity, peaked about 2,000 years ago, then the stones of the pyramids made three thousand years earlier were in a slightly less pronounced state of hardness and perhaps adjusted more easily to each other. These hypotheses are no crazier than the more or less official ones presented by science. Has not the earth oscillated between periods of heat and glaciation? Have not the Earth's poles and magnetic fields also changed their locations? Don't the planets manifest themselves in different states of hardness? What is the meaning of the rings of Saturn?

Sympathies and Antipathies

Our inability to understand the appearance of new diseases in agriculture at least has the merit of shaking up certain convictions. Those of us willing to admit our ignorance and open up our spirits are more and more numerous. These dramas have brought into relief the limitations of a strictly materialistic approach which, if it is accurate, always remains incomplete. It is comparable to a musician presenting an audience sheets of his written music, but abstaining from playing it. Notes of music on paper are only an abstract skeleton that is devoid of charm. Chemistry also opens up new perspectives, if one considers how the sympathies or the antipathies of the elements toward each other are the echo of that far-distant genesis of a world in which the laws of gravity expressed themselves less powerfully, where separations were less pronounced. Experiments are actually being conducted in space, where the gravitational field is not as strong, to observe the changes in the way that chemical reactions behave. Observing the animal or vegetable worlds can also help us to interrogate ourselves about these special affinities that each species has for infinitely small doses of substances diluted in the water or the air. Within a few months a cuttlefish builds up a backbone of several ounces of calcium out of a concentration of only trace amounts of calcium in seawater. How does it do this? From what source does each plant draw the ability to give birth to a flower, the forms, colors and odors of which are always unique and specific?

There are six planets that are of particular interest to winegrowers who have chosen to work with biodynamics: Saturn and Mercury linked to the forces of heat, Jupiter and Venus to the forces of light, and Mars and the Moon to the forces of water. Let us leave out of the discussion the three planets which are the furthest away: Uranus, Neptune and Pluto. The positions of these six planets in respect to each other and the earth are always evolving. Their respective situations never repeat themselves in exactly the same way. Their course around the sun is always a bit different — their rotations are not perfect circles, but rather, ellipses, and an ellipse has two "centers." The sun is obviously one, but what is the other? Moreover, this system it is a source of diversity which is quite compatible with life. The angles formed by the positions of these planets in relation to the earth can stimulate for one day sometimes a particular quality and underline "water," "light" or "heat" aspects in a different manner, according to whether it is an internal planet with a year shorter than ours or an external planet with a year longer than ours.

We refer to situations which are visible from the earth. This is called geocentrism and postulates that the earth is the center of the universe. A plant, which has no brain, of course, certainly has a geocentric view. It does not know that the earth is turning around the sun — for it, the sun rises in the east and later sets in the west. With all this in mind, let's see how we can benefit more from heat forces. When two "heat" planets are diagonally opposite the earth, so that the earth is positioned between them on the same straight line, their effects are reinforced. This situation is called an "opposition." When it occurs one reinforces the prevailing effects by utilizing dynamization, or even by scratching the soil. Scratching a soil is a little like opening the curtains of a room. If it is raining outside, or snowing, opening the curtains or shutters allows those inside to experience more of the outer situation. These practices permit the living elements in the soil to better impregnate themselves with these forces. Whether this agricultural activity be a dynamization applied to the leaves or a hoeing of the soil, the principle remains the same. When it is a question of planets linked to the influence of water, the same work concentrates the force of the plant in its leaves. This could reinforce the quality of the leaves of vegetables like lettuce, cabbage, etc.

On the other hand, when two planets are on the same line, both on the same side of the earth, this situation is called a "conjunction." In this situation the effects of the planets are in conflict, and dynamization or work on the soil is not recommended. In other words, in addition to knowledge about positive situations that you want to reinforce, you also need to recognize negative conditions that should not be encouraged.

These phenomena have been observed during very numerous experiments. In the infinite diversity of this "ballet" of planets and the earth around the sun, some aspects are beneficial and constructive, while others are negative. The most favorable angles for an act of reinforcement are, for example, 180 degrees for the "opposition" and 60 degrees for a "trigon" or "trine" in which planets and earth form an equilateral triangle. The disturbing effects are around 0 degrees (conjunction), 90 degrees, etc. Following these principles, these phenomena are amplified from generation to generation for cultivated seed. The influence of the planets can be physically measured in your garden and in your home if the rules of life express themselves there appropriately. If you use genetic seeds on a soil which has been deafened by weed-killers and chemicals, you take the risk of wasting your time. Divide a garden into two sections. In the first section the earth is worked on a "water" date before planting wheat. A few days later, and no later than that, the experiment is continued on a "heat" day, planting the same batch of seed in the second section that was used in the first. To take the experiment further, it is possible to proceed simultaneously with a dynamization. The results are substantial. In the section where the forces of water were solicited, the growth of the stalk was reinforced. In the harvest there is going to be a lot of straw, but not much wheat. On the second section, it was the forces of heat and fruition which have been invoked The stalk is smaller, but the head is much larger, because the growth of seed has been stimulated. Our intervention has been in the same sense as the direction of the plant.

In her books (particularly *Work on the Land and the Constellations*), Maria Thun follows up on her experiments over very long periods of time by replanting the same seeds, which are charged with a "quality" and a particular orientation. After a few years, the wheat which was not planted under the sign of heat no longer gave any harvest unless assisted by chemicals and was destroyed by disease. On the other hand, the section on which the forces of fruition were accentuated was much more resistant to disease and had high yields at harvest. The differences can be quite important, on the order of 25 to 30 percent, according to the harvests. The same experiment done with strawberries was particularly eloquent at the moment of a blindfold tasting of the fruit. On the water date harvests, in spite of a mature appearance, they didn't taste fully ripe. It was the leaves that were strengthened and this showed up in the taste of the fruit.

This presents a practical problem — when to act and how many times? All good winegrowers know that days of rest are few and far between, and it is awkward to adjust one's work day to the demands of the sky and the caprices of the weather. It is fortunate that the moon goes around the world every 28 days. The rapidity of this rotation frequently permits us to recognize aspects propitious to the vines. The moon acts as the mirror of a situation that is further away. All this is true in a context where life can express itself, where the receptive properties of the soil are active and where the very nature of the vine has not been tampered with. These examples underline yet again that behind each planetary aspect hides a force that we can choose to strengthen or to weaken in our vines. But there is nothing like real-life experience to help make up your mind. Before definitely closing the door on a different way of looking at agriculture and wine growing, give it a try. You take only one risk: it might convince you.

Of course some will always prefer and continue to use chemical fertilizers to improve their harvests, working against the laws of nature, without worrying about what is behind this glittering delusion. If we do not again find the path to certain fundamental truths, we will soon fulfill the predictions of the World Health Organization mentioned in a recently published book (*Les poubelles dans nos assiettes*, by Fabien Perucca and Gérard Pouradier), one person in four will be stricken with cancer within four years and one person in two by the year 2010.

In not considering the energetic dynamics of our daily lives (in our environment, food, etc.), we have reached a stage where our health "capital" has declined to a dangerously low level. We have never seen so many legal measures on the cleanness of food, or so much concern over viruses, pathogens, etc. At the same time, our food supply has never been so empty energetically and therefore so tempting to all these destructive factors.

Astronomy is not Astrology

Let us not confuse astronomy, to which we refer ourselves in this chapter, with astrology. The study of astrology is based on a circle divided into 12 equal parts which represent the 12 months of the year. In his book, *Jardiner avec la lune*, Xavier Florin explains it very clearly. An astrological sign is determined according to the position of the sun in this symbolic circle when a child is born. At the beginning of our era this circle accurately represented the locations of the stars in the sky. However, this position has changed due to the precession of the equinoxes, but the astrological circle remains fixed. This phenomenon is easy to understand. Every 72 years the stellar background as seen from the earth moves one degree in relation to the sun. Thus, if you were born at the beginning of spring, your constellation according to astrology is Aries the ram, and according to astronomy it is Pisces. There is a difference of 30 degrees over a period of 2,160 years. Under the Romans, in the spring the sun did not rise in Pisces, but in Aries. Under the Egyptians, it was in Taurus, etc. The sun makes a complete cycle in 25,920 years, which is what is called the Platonic Year. Much can be learned about this cycle by reading the book of Walter Cloos, *The Living Earth.*

In summary, let us remember that the stellar background, as seen from the earth, is not fixed and that the symbolic circle designed by astrology has not been revised for the last two thousand years (to see more clearly this difference between astronomy and astrology, turn to page 155). For the vegetation, which does not think, mainly astronomy is chiefly of interest — that is to say, the real situation of the sky the plants are growing under. If one takes into account the stellar and planetary systems, the earth each day is a reality which never repeats itself in the same way. Many parameters participate in the diversity of this complex living universe, of which we know only a tiny bit of the movements, without so much as understanding it to be a creative organism. And for that it would be sufficient to try to manifest the effects. Do you know of a single research institute which is studying this subject? The risk of discovering free fertilizer all around us, qualitative and non-polluting, is undoubtedly too great.

Some people probably jumped at the words "Aries" and "Pisces." The weak capital of credibility that we have been trying to establish is seriously drained. But it is too bad not to go beyond a vision that is a caricature, if not defiled by the excesses of a commercial and industrial use of astrology. For a winegrower, the essential information to know is that the year is divided into 12 different "qualities" of the sun. To make myself clear, when seen from the earth, the position of the sun at the same time each day is different from month to month. Instead of referring to the January or February sun, the old ones used the name of the constellation before which the sun was situated. They spoke of the sun in Capricorn or the sun in Aquarius, etc. But how to explain the choice of these symbols? Frit Julius has been interested in this question and has tried to translate this language in his book, *The Animals of the Zodiac*. His analysis is well thought out. Each animal incarnates the particular quality of the sun during a precise period of the year. Everyone knows there are appropriate dates for planting each different species. Spring wheat and autumn wheat have planting times that are in dramatic opposition and, if they are not respected, the harvest will be nonexistent. In a natural context the snowdrop does not flower in May or June but, as its name implies, at the end of January or beginning of February. There are thus many particular qualities developed by the sun at different times of the year that awaken the growth of plants.

At the beginning of the year, the sun is in front of the constellation of Capricorn. This animal is the ibex. If you try to understand it, it has some interesting characteristics. In the middle of winter, motivated by an incredible determination, it remains on the mountain heights in spite of the tempests. It refuses the relative comfort and protection of the forest even though it has access to the forest. It has a thirst for the heights. One might say that it nourishes itself on them since its retreat imposes a fast that lasts for several weeks. In spite of its somewhat bulky stature, its agility is surprising. Sure-footed, it knows how to make use of the slightest bumps on the smooth and almost vertical rock face which it inhabits. The ibex dominates its heaviness, ruled by gravity. This animal is a very judicious choice to represent the first rising sun of the year which permits the days to grow longer, vegetation to progressively return through the attraction of the new sun and our spirits to take off toward the charms of spring and summer. It is a considerable change for life forms on earth to see from Christmas onward the days getting longer. During this winter period, man also draws on his reserves of will power. He wants to go toward the sun and the light. If we live out the seasons as humans of the earth in the most profound depths of our being, it is

Astrology & Astronomy
(Northern Hemisphere)

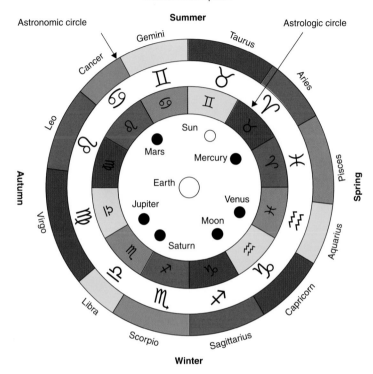

Astronomic circle

Astrologic circle

Northern Hemisphere

Summer

Autumn

Spring

Winter

Astrology: *circle divided into 12 equal parts (the inside circle on the diagram).*

Astronomy: *circle whose division corresponds with the real positions of the constellations at the present time (the outside circle on the diagram).*

As we can see, the two circles no longer coincide. Biodynamics only takes into account astronomy.

The sun, moon and planets move through this circle according to their respective orbital periods — a year for the sun, 28 days for the moon.

The "tendency" of the planet, joined to that of the constellation before which it is passing, can be stimulated in a plant by dynamization or a method of cultivating.

The dominant forces in the root are the forces of earth.

The dominant forces in the leaf are the forces of water.

The dominant forces in the flower are the forces of light.

The dominant forces in the fruit are the forces of heat.

Zodiac	Planets	Forces	Plants
Violet	*Earth*	*Earth*	*Root*
Blue	*Moon, Mars*	*Water*	*Leaf*
Yellow	*Venus, Jupiter*	*Light*	*Flower*
Red	*Mercury, Saturn*	*Heat*	*Seed, Fruit*

The Planets   155

Astronomy & Agriculture
The three positions of the moon in the constellations of heat

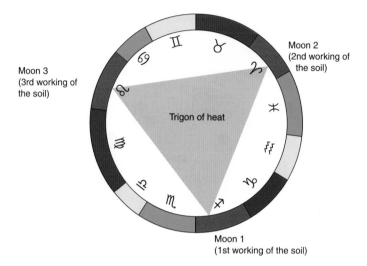

Moon 3
(3rd working of
the soil)

Moon 2
(2nd working of
the soil)

Trigon of heat

Moon 1
(1st working of the soil)

The most favorable moments for the vine are those associated with the forces of heat, which are propitious for fruition.

These forces emanate from the constellations of Aries, Leo and Sagittarius, which form an equilateral triangle between them, called a "trigon." These forces emanate also from the planets Mercury and Saturn.

To treat with a trigon means to carry out a dynamization or agricultural activity on the vines, for example when the moon, our satellite, passes in front of these constellations (see diagram).

A theoretic example of planetary disposition in which cultivation methods are very beneficial to the vine:

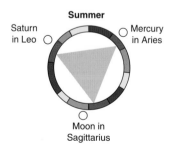

Summer

Saturn
in Leo

Mercury
in Aries

Moon in
Sagittarius

unthinkable to remain insensitive to the choice of this animal and its will to live as the earth does during this period of darkness.

Let's take another example. At the beginning of spring, the sun is in the constellation of Aries, the ram. Why this animal? The equinox has passed. The days are longer than the nights. The vegetative forces are fully externalized toward the visible world. Have you ever witnessed a combat between rams? The animals back off deliberately and face each other without being in a hurry and without spite. The combat seems to be dominated by a superior force and the fighters show no impatience as they await the signal to start. Then with all their might, they powerfully charge each other. The shock is muffled and violent. They pause for a few moments, get up, back off and renew their combat, which will not stop until one of the two remains lying down. It is a frank and honest fight, without low blows, very different from a fight between dogs, for example. The rams are then carried along by a constant and formidably powerful external force. The difference with the preceding example is that this will-power is not internalized in an immobility which permits it to confront the snow and cold, but is expressed in movement. Their determination is implacably turned toward the external, into space, thus toward matter. It is the symbol of the manifestation of the forces of spring. These forces awaken our soils. When conditions are favorable for the vine to emerge from its winter sleep, it grows several centimeters each day. The translation of the language of the animals of the zodiac provides a new way of looking at the meaning of astronomy. One can then progressively read in it the wisdom of symbols chosen long ago.

The four states of matter also express themselves in the 12 constellations of stars that the sun (as seen from earth) travels through during the course of a year. Over a period of 28 days. the moon does the same. The constellations of Aries, Leo and Sagittarius are linked to the state of heat (fruit). The light (flower) is reinforced by the constellations of Aquarius, Gemini and Libra. The watery state (leaf) benefits from the forces of the constellations of Cancer, Pisces and Scorpio. And the mineral state (root) is favored by Capricorn, Taurus and Virgo. In each case, the three constellations form an equilateral triangle between them. By referring to an appropriate calendar, one can benefit substantially from the passage of two heat planets through a heat constellation, or the passage of the sun or of the moon (again, as seen from earth) in these constellations. When they form a particular angle with the earth, these influences are again accentuated (see for example the "heat trigon" on page 156). A healthy agriculture is receptive to these planetary situations. The specifics of these propitious moments are reinforced by appropriate cultivating activities — working the earth, planting, dynamizing and harvesting. On the contrary, if a man has isolated a plant from its living context, it is almost totally insensitive.

The sun develops three qualities of heat, according to its position in the sky. The sun in Aries manifests the forces of spring of which we have already spoken. When the sun enters Leo, its descent in the sky has already begun. This movement toward the earth responds to the desire of the vine to internalize, which slows down the growth of its shoots in order to concentrate on its fruit. The forces of fruition stimulate the formation of seed. Those working in nurseries should learn to court this sign, whose influence induces an acceleration of the germinating process and an improvement in quality. By working the earth in the sign of Leo, winegrowers help the vine to reinforce the growth of the seeds, which bring natural tannin to the vine. The sun in Leo is then favorable to the maturing (aoûtement) of the shoots and the ripening of the grapes. Sagittarius is a sign of winter heat. The descent of the sun in the sky is nearly complete. The days are no longer getting shorter. The turn of the cycle toward a rising sun is on the way. The arrow in the drawn bow of Sagittarius is pointing upward to underline it. These backgrounds are not exclusively seasonal. We find them thanks to the movements of the moon and the other planets. Their effects are slightly modified in relation to those of the sun, but the principles remain the same. A moon in Sagittarius at harvest time is not a bearer of all the forces of winter. The fermentations will however be more calm, while they will be more tormented if the harvest is brought in during a moon in Aries.

Aside from agricultural activities, the winegrower can orient his wine toward one or the other of these planetary and stellar aspects. The combinations are multiple. This diversity gives a man the full measure of his role as conductor of the orchestra. It is up to him to give expression to the music he wants us to listen to. But let us repeat that biodynamics can be effectively practiced without worrying about these planetary or stellar realities. The panorama being more restrained, the results may simply be less profound. Isn't the winegrower who delves into the complexity of these exchanges to elaborate a wine complicating his task a bit? Undoubtedly, but he enriches his mission by approaching it as an artist. It was, by the way, not without good reason that the true peasants who had received this knowledge as an inheritance, without being able to explain its origins, greeted the false progress of agriculture with such distrust. They detected the lies contained in speeches that were too optimistic to be honest. However, they had no way of defending themselves against the representatives of modern agriculture or against their mockery and ignorance of the profundity of the laws of life and the complementarity of the animal and vegetative realms. This intimate knowledge of their environment lost its foundations from one generation to the next. It started primarily with World War I, in which millions of true farmers were killed. Perhaps if they had survived, the massive use of chemicals in agriculture would never have become the norm. Today part of the ancient instinct is all that remains intact — the full knowledge is gone. A farmer knows that he is right without necessarily knowing how to explain why. The other celebrated over results whose indirect costs he had no idea of. Two opposite worlds confronted each other. In *La chasse au bonheur*, Jean Giono, perhaps the best French spokesman for the soul of the farmer, described them magnificently, particularly in the chapter entitled "The Zodiac." If biodynamics has a mission to accomplish, it is without a doubt to reconcile these two attitudes which are true types of complementary thought. It can't be done without presenting physical evidence of the effectiveness of biodynamics. Progress is a necessity, so long as it serves what we have acquired and is based on the resources of our environment. Why look elsewhere, when it is free and within arm's reach? The Research Institute for Biological Agriculture near Basel, in collaboration with the Swiss Federal Office of Agriculture, has been carrying out a comparative study since 1978 of three systems of agriculture: conventional, organic and biodynamic. Far from classifying biodynamics among the trivia that go in and out of style, this Institute has been intensely interested in it for 20 years.

In summary, biological methods resulted in harvests that were 15-18 percent less than with chemical methods. The two organic procedures used the nutritious elements of the soil and energy much more effectively. The biodynamic areas have the highest biomass (amount of living matter in the soil) and more extensive mycorrhizal activity in the roots of the plants. Aren't those ideal conditions for bringing into manifestation one's label of origin? Unfortunately the study does not take into account the prohibitive expense of the secondary effects of conventional chemical agriculture. Here we should point out that the Dutch government supports biodynamic farmers with a very substantial per-hectare subsidy.

The images below compare the roots of wheat grown with conventional and with organic or biodynamic methods. Similar results could be achieved with vines. Can't we call this a better expression of the terroir effect, a better intimacy between the root and its soil?

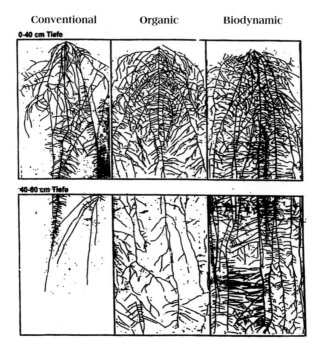

Chapter IX
Understanding the New Qualitative Tests

Biodynamics, homeopathy, and the debates over the weakening effects of cloning and genetic manipulation participate in a global reflection whose point of departure is a quality — a creative energy that is inaccessible to our physical senses. These intangible phenomena are neglected in spite of their physical manifestations. These qualities can from now on be brought into relief by new scientific tests which capture the creative momentum in an image. These images give us information on the ability of a food product to give humans nutritional energy.

Their interpretation, contrary to their appearances, requires long practice. Nevertheless, the neophyte can then and there detect these appreciable differences, which constitute a primary point of reference. There are two traps to be avoided. The first consists of seeing something where there isn't anything, the other is not seeing anything where there is something. For the sake of clarity, one must begin without preconceptions of any kind. These tests open the way to new knowledge to the extent that they neglect the quantitative approach in favor of qualitative criteria. These two approaches were summed up by one of the representatives of the Waldorf School (see Appendix III). There are two possible ways of getting to know a child, you can run a blood test on him or you can ask him to draw a picture.

Before the appearance of sensitive crystallization and capillary tests, some experiments had permitted the vital properties of different substances to manifest. Thus Haushka, the author of *Nature of Substance*, tested the energetic qualities of straw, wood, gas, coal and electricity.

The protocol was simple. It was a question of boiling a given quantity of water by using different fuels. The water was then used for sprouting seeds. During his lecture on the results of his experiment, Haushka was not satisfied with quantitative criteria (weight, height, etc.). He was interested in the proportions of the roots, the leaves, the stalks. Substantial differences were recorded. Water is receptive to different qualities of energy, whose specifics are manifested at the level of vegetation. The last adepts of heating with wood know to what extent at equal heat this energy is different from that furnished by electricity. This subtlety of the heat emitted by a wood fire can never be registered on any thermometer. During a second experiment, Haushka boiled water in containers made of different metals: gold, silver, copper, iron and tin. (Each metal has a particular affinity with a planet: gold for the sun, silver for the moon, copper for Venus, iron for Mars, tin for Jupiter, and lead for Saturn. These links and their effects are explained in the book by Dr. Bott, *Santé et Maladies*). According to the same principle, Haushka used these batches of water to sprout seeds. Gold favored the most balanced growth. Thus one sees that an inventive spirit can always try to bring into relief differences that are hardly noticeable at first glance. The danger is to concentrate one's observations on isolated or congealed elements. That is one of the pitfalls in agricultural research, in which specimens removed from their environment are imprisoned by computers, thus by a binary system incapable of seizing the living world.

Sensitive Crystallization

When Rudolf Steiner was questioned by Ehrenfried Pfeiffer about ways of bringing into relief living backgrounds, Steiner explained the principles of sensitive crystallization. The basic idea is simple. A solution of copper chloride is mixed with an extract (juice or maceration) of a plant. The product obtained is poured on a Petri dish and allowed to crystallize at a predetermined temperature and humidity. The solution of copper salt modifies the rigidity of the metal and the heat develops the subtle qualities of the substance which is going to be able to manifest in a sort of weightlessness, thus somehow delivered from gravitation. The crystallization manifests the characteristics of the formative process. If the substance analyzed is dead on the level of living forces, the copper chloride dries leaving a sort of stain. In the opposite case, the living forces organize the copper chloride and form an image comparable to the ice crystals on our windshields after a hard freeze. The image is then analyzed according to different criteria such as its general aspect, regularity, fineness, the clarity of the center and the peripheral forces.

The two images of crystallization of hawthorn and a pill of hawthorn (see below and following page) need no comment since the differences are so obvious. The hawthorn pill is evidently amputated from the living forces of the plant. This is most probably due to the manufacturing process, which probably overheated the substance. A study based on quantitative criteria would not show any difference. Here we measure the usefulness of these tests, as much for producers as consumers. They allow us to avoid incriminating the basic substance as ineffective afterwards. In homeopathy, where the elaboration of dilutions is submitted to very subtle forces, these tests can constitute a valuable supplement of information.

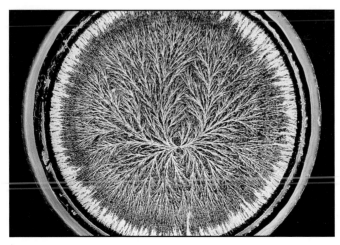

A crystallization of hawthorn characterized by a very clear structure.

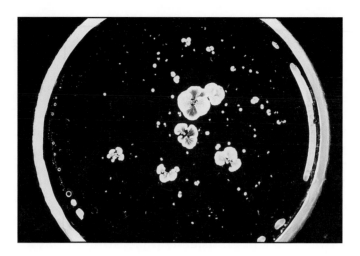

The crystallization of a hawthorn pill, as commercially sold, which no longer has any living properties. The vital force of the substance was thus not able to influence the copper chloride. The manufacturing process was done at too high a temperature.

Images of a crystallograph done on human milk, destined for infants.

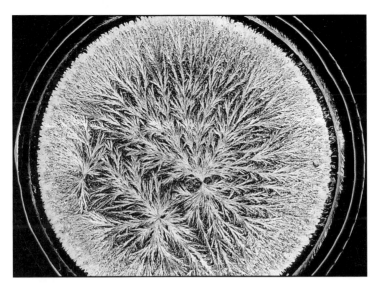

Powdered milk which, as recommended by many pediatricians, has been fed to generations of infants. One sees that the vital forces of a product are not necessarily linked to the matter which it is composed of.

In Germany the crystallization are used more and more frequently in the context of human blood analysis. They permit the detection of a potential illness before the appearance of the first physical symptoms. Blood always carries an imbalance before it materializes into an illness. Crystallography is also used to test the effectiveness of a therapy and the progress of its effects on the sickness under treatment. By comparing the images, the doctor can quickly adjust the medication as needed since the crystallization allow him to see the appearance of any side effects and to identify any progress that has been accomplished.

With the friendly cooperation of Colonel Gauthier, responsible for civil protection in the Maine-et-Loire department, we treated potatoes with gamma rays before submitting them to tests of sensitive crystallization. At a glance we could measure the perverse effects of such a treatment on agricultural food products. Exposing food to gamma rays, which is no longer illegal in Europe, brings about a drastic energetic change. Gamma rays, like herbicides, destroy everything they touch — the living elements, the microbes and bacteria. Should we destroy what is pathogenic

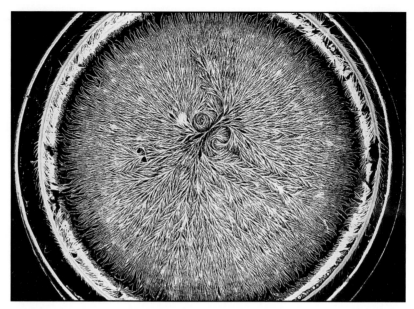

Crystallographic image of biodynamic white wine.

Ordinary white wine. These images are so explicit that no comment is necessary.

The preceding images were produced by the L.A.P.A.T. (Laboratoire associatif pour l'application des tests sensibles), with the help of Bernard Prieur and Marie-Françoise Tesson.

without taking into account what is beneficial? This censorship of the forces of life also has the commercial advantage of delaying the appearance of rot. The consumer fills himself with food which, having lost its living structure, has become carcinogenic.

Knowing that these tests exist, we have the right to ask how the ecological organizations could have been satisfied at the time when nuclear testing was resumed with a simple physical monitoring of pollution? Physical radioactivity is only one of the aspects of the consequences of a nuclear explosion whose effects on the living organism of our planet we are trying to ignore. Maybe the truth is not nice to look at. One has the image of life that escapes, leaving death behind. Biodynamics is the exact opposite, since it concentrates the life forces in their process of incarnating on earth. Besides it has a substantial impact on radioactivity. At the time of Chernobyl, measurements were made on a collective farm in Poland where Maria Thun was conducting biodynamic experiments, and they showed radioactivity reduced by 90 percent!

These tests are extremely sensitive and should be done with the utmost care. The same wine may show results in the evening that are quite different from those obtained in the morning. Life is like that. Sometimes other quality tests need to be performed at the same time. One organization in Denmark has developed a computer program to interpret these pictures, of which they currently have more than 30,000. Their workload is enormous, as more and more people begin to search for a deeper understanding of what is called quality.

Capillary Tests

Capillary tests are also called morphochromatographs, which are not to be confused with the simple chromatographs whose objective is to separate a compound into its constituents. On the contrary, morphochromatographs study the reactions of these constituents when they are reunited under a precise experimental protocol. The basic principle is still founded on the capillary migration of organic substances (mixed with a solution of metallic salts) on filter paper. Depending on the salt or salts used, the components of the substance develop different shapes or colors. The capillarity of paper permits the liberation of the product under analysis from the laws of weight (gravitational forces), which are often an obstacle to the manifestation of the qualitative properties. In brief, two reactions are possible: either the juice does not oppose the rising of the reactive agent, or it blocks the rising and no shape develops. Between these two extreme situations, there are evidently numerous variations.

This very sensitive method is more delicate to interpret for a beginner. One cannot speak of crystallographies or morphochromatographs without mentioning Dr. Kolisko's remarkable book, *Agriculture of Tomorrow*, which provides us with many images and comments on these tests. In the context of wine growing, these new tests, especially the sensitive crystallization, have almost not yet been applied. This remains virgin territory with the exception of a rapid study that tried to explain the causes of "golden-yellow" disease. However, the subjects of experiments are varied and could provide valuable images of sterilizing filtrations, sulfur, cold treatments, the quality of biodynamic preparations, the effectiveness of dynamizations, the analysis of the soils and their affinities with the vegetation. It would be sufficient to take a wee bit of the subsidies devoted to agricultural food research and put it toward study of these alternative methods, but the organizations involved prefer to confine themselves to a quantitative approach, as it is clearly more profitable.

In his book *Sensitive Chaos*, Theodore Schwenk, already quoted, also refers to tests of a sensitive drop which have similar objectives and tests of the sensitivity of flame. All this underlines the great sensitivity of the world that surrounds us.

Demeter, which is the world trademark for biodynamics, sometimes requires the crystallization of the products it endorses. Let us point out that the Demeter label is granted when the entirety of an agricultural domain has been biodynamic for three full years.

It is important to mention here some work presently being done in Germany (the researchers wish to remain anonymous at this stage) on the germination of seeds exposed to different notes of music. The differences in growth are striking. To understand this more fully, keep in mind that music is based on seven notes, which are linked to the seven planets. They are made of specific rhythms, and music brings about a rhythmic order, unless it is misused. This may also help to explain the healing forces at work in music therapy.

Wouldn't it be interesting to see the effect on taste if some specific musical notes were played for the vine? All plants receive forces from all the planets, of course, but one is always dominant. For the vine, Mercury is most probably the one. Remember that in Greek civilization Mercury was the god of merchants, but also of thieves — thus, a force dominated by movements, exchanges, robberies (although unpleasant, robberies still contribute to the circulation of goods).

Exposing vines to one note or another at regular intervals, played by a real musical instrument, would be most interesting. Which instrument would be the most appropriate? One made of copper (Venus forces), of silver (moon forces), or of wood? If the last, then which wood? Trees can also be dominated by one planet or another (in the Goetheanum in Basel, Switzerland, Steiner used seven columns made of seven different types of wood). Should one use a wind or a string instrument? All these questions remind us of how farming can lead us to knowledge that would help us become the "artists of the earth" — and winegrowers can lead the way.

Chapter X
Life Is Energy

When one talks about biodynamics — or the "quality forces," which quite clearly produce real quality in the grapes — one always refers to energy. What one has to understand is that living matter can only take shape in the physical world by impulses of energy. These energy impulses come, as we have already seen, from the solar system — and not just from the sun — as well as the larger stellar system, which interacts continually with Earth and our solar system. To believe that something situated millions of kilometers from the Earth has no effect because "it is too far away" is a denial of a whole system of communication and exchange, a system that we use today — if somewhat poorly, because we have so far only partially identified it.

This system is active in all that surrounds us — without it the slightest "chemical reaction" could not occur, because the sympathies and antipathies of molecules are linked to it. All this is experienced by every living organism on Earth, these organisms that receive life from and whose behavior is determined by the originality, the unique form and function — or to put it more visually, by the "small part of the global life" — that they are linked to by their configuration. As we have already mentioned, it's like individual notes, which, once put together, participate in the melody that is life on Earth. This is very important to understand in order to avoid errors, such as we will mention later on, in the newly burgeoning use in our society of energetics.

This global comprehension of life can help to understand how an eclipse can be "felt" by plants sometimes a day or more ahead of time, or how a dog can predict an earthquake. In this "flow" which is time, nothing is isolated or separated. Everything is mingled — the past acts in the future, the future affects the past, as in a painting where each color highlights the effect of the others.

It is also through an understanding of this life system that each particular planetary situation (opposition, etc.) can be used to reinforce a precise impulse in a vine if one so wishes.

As soon as one talks of living matter, one inevitably invokes the notion of death, because life and living matter are opposed polarities. The more one digs into living matter, the more life is extinguished to leave behind a harsh, immobile, mineral world. But death can only exist on a physical, material level, not on the energetic level, because is not the physical that makes life — on the contrary, it is life which gives form to the physical side. The physical is a world of life rendered more dense by gravitational forces. All this latent world, in or behind the matter that surrounds us, is made up of frequencies, therefore energy, of rhythms that sculpt the matter, that dominate it completely, lifting it up to give it each time a different form.

Finally, one finds in this understanding a defense of Goethe, who mentions the "idea" that is behind a plant. All terrestrial realizations can be understood, he says, as an idea — one could translate this into today's language as "a number of frequencies," like data — each time specific and belonging to a whole. This understanding is essential to quality agriculture. As soon as one understands that physical emptiness is not void of energy (which physicists have proved over the past few years), one can begin to understand the "other side" of the physical world, the realm of the "ideas" or "archetypal forces," the diagrams, so to speak, which appear physically in all living organisms.

And here one must not consider only the elements that transform "idea" into matter (with the genes acting as "interpreters" linking the two worlds), but also the harmony and complementarity of all these ideas together. Its what the elders of different civilizations (Kepler said it as well) called the music of spheres, or harmony of worlds. One mustn't act only on the system (in this case, the genes) that transmits these ideas to the physical world, or to the system that submits the "idea" to the needs of the physical world. By limiting our vision in this way — with methods that are unbelievably violent or barbaric, such as the nuclear gamma ray — one destroys the "coherence" of the idea itself, its architectural plan and its link to other "ideas." In other words, in addition to the specific gene, one also modifies the forces of unity which are linked to these genes, or are latent behind them. By destroying the unity of the idea by a partial modification, then, one has created chaos or a carcinogenic process, that is, a proliferation without structure, without an organizational force, without coherence. It is somewhat comparable to words from a poem printed at random.

To make humankind eat food that comes from this disorganization is a serious act. When an animal is submitted to such treatments, it can only answer by leukemia. To prevent leukemia in turn would lead to another response just as serious, and so it must be until one reestablishes the coherence of the "idea" that is latent behind every species. One could go further as far as man's health is concerned and talk about hundreds of illnesses labeled "genetic" which suggests to a neophyte that they are simply due to the disorganization of genes — like a typo in a text — that could occur to anyone. This is an illusion made possible by an incomplete understanding of the laws of life. One has to recognize that each of our actions interacts — as in a form of resonance — with the energetic systems that look after the Earth and keep its inhabitants alive. What one calls "chance" is only lack of knowledge. The hereditary current is only support, in which the individual current, which is talked about too rarely, finds or doesn't find fertile ground in which to blossom. There are many families with several generations of musicians (hereditary current), for example, but there has only been one Johann Sebastian Bach (individual current). This is another subject, but it is important to recognize, because all of this results in the understanding of our individual capacities, those with which the winegrower makes his wine.

Matter, then, is ultimately only a point of reference for our senses. To try to reach new quality limits on Earth without understanding the matrix of life-forces only creates an imbalance, with a debt that will weigh for a long time upon human health and creativity as well as future generations. A person out of tune with his food and with these "frequencies of life" that created him will become ever more isolated, and the world becomes limited to the physical perceptions. The forces that he should use with fidelity and creativity to make use of the world into which he was born (and whose full perception he left in the first few months of his life) might end up being used in an inappropriate manner. His attitude is contrary to the wisdom generated by a receptiveness to the laws of the universe.

We can now better understand the issues of a healthy agriculture or, even better, a biodynamic agriculture, which, when properly carried out, intensifies within the plant these celestial rhythms and their curative virtues. Along the same lines lies an understanding of the building materials, forms and locations used by the old builders (we have already talked about the cellar and the barrels) to create dwellings that concentrate the forces of life.

In a book on biodynamics and wine, it is important to open doors, to give to the younger generation of farmers, a generation that is "stupefied with a dry knowledge," lacking inspiration, an opening to these potentials. This is not a course on religion (from the Latin *religare,* meaning "to link"), but a course in truly modern sciences. With this approach one can find the qualities of a "green thumb," or a reconnection with Earth. Thus, through one's own creativity, and by moving away from the uniformity imposed upon us, one can attempt to become an artist of the Earth, and a healer, as well.

Science in the past has always dealt with the palpable physical world, and has always neglected what sculpts it and puts it in place — rather like a food critic attempting to understand a dish without taking into account the gestures that the cook uses in its preparation. Modern-day man has isolated himself through his intellect and his ego, his "I," which is given to him as a source of autonomy. He now must learn to act *correctly* on the world around him by his gestures and his thoughts, and this type of creativity is also what biodynamics needs to express itself fully.

In the old days mankind was less individualistic, more community oriented, and thus had hardly any conscious understanding of this universal system. Rather like a newborn child; it was almost by intuition that he felt where the best place was to built a house, found which plants could cure illness, felt that agriculture was like a link within a vast system, worshiped stars, etc. He knew he was right, but could not say why. Animals have this instinct, which supports them and identifies the things they need and the things they should avoid. They do not have a "choice" — they follow their "instinct," an unconscious link to the quality world which generates life. It is through this link that the beaver has the know-how to build a dam, or the bee knows how to recognize a flower that has nectar.

This collective feeling/intuition is what enabled past societies to be organized upon very precise lines of conduct. Little by little, man learned to use these energetic laws to modify the physical world — we have already talked about the pyramids or the temple where man cured the sick, etc. Sometimes he learned how to link his thoughts with precise forces to such an extent that he managed to act on the complex structures that generate living matter. From there is born the notion of white or black magic, or more simply, of alchemy — a subject which is currently being researched by the CNRS (French National Centre of Scientific Research). It is in this way that Steiner's title *The Science of the Occult* should be understood. It is about broadening the *basis* of science, to make it branch out further with the aim of improving the health of mankind and the Earth.

The physical concept of the world is extraordinarily destructive for the forces of life, therefore for mankind's health, and of course for the full expression of this quality world that one seeks out in wine. We only admit the visible and do not measure the energetic level, for fear of questioning the very basis of our knowledge. We have even reached the point where homeopathic products — which get their powers from the energetic world and can be produced extremely cheaply (but are not patentable) — are banned.

Now we can better understand that biodynamics is simply an "energetic" use of all these life processes, which nourish in an invisible but real way the plant and the earth, rather like good news (without weight or volume) that will "carry" us through the day. In our physical acts, biodynamics brings resonance with the archetypal ideas of the plants. The more one tries to understand the deeper sense of the biodynamic preparations (U.J. Koenig's booklet *The Biodynamic Preparations,* and the writings of Livegoed dedicated to biodynamics possibly go farthest in this direction), the more one has to go back to the secret laws that preside over life on Earth.

Finally, we have seen that mankind today, through his lack of understanding, isolates the Earth from its energetic context (the energy that gives him life every day), and the isolation curve is exponential. The result is that the magnetic fields of the Earth are changing and diminishing (for convincing evidence compare aeronautical maps from 10 years ago); they are also becoming unpredictable — pilots will confirm this. The consequences are clear as far as the weather and climatology are concerned. The Earth's magnetic North is moving faster and faster (from 1984 to 1994 it moved as much as it had from 1862 to 1984, and its movement from 1994 to 2001 surpassed even the 1984-1994 decade); volcanic activity has also been multiplied by four since 1973, and the rate is accelerating.

The same applies to earthquakes. If one understands the Earth as a living organism, one reads these as reactions to an obvious exhausted planetary state, like a jumpstart, attempts to fend off the asphyxiation of the forces of life. To limit earthquakes to continental plates which come together is like limiting a smile to the muscles pulling the lips. It seems as though many want to put the human being to sleep in the name of abstract cycles or unavoidable things over which we have no power. In reality, the situation is the opposite. Each individual has his role to play in fighting these excesses which result in the acceleration of serious weakness and illness (physical or psychological, terrorism included) in mankind and all living organisms. Comprehension is mediocre and healing on the physical side is becoming expensive and inefficient.

The symptoms for a major change on Earth in the near future are here: the magnetic field wasting away, the changing of position of the magnetic North pole, increased numbers of earthquakes, volcanic eruptions and climatic disorders, including the slowing down of the Gulf Stream and maybe even its halting in less than 10 years. When we are exhausted, we sleep. How can an exhausted earth sleep? The answer is simple: by inverting its polarities. In other words, the minus becomes a plus, the North Pole becomes the South. This is a sort of rebirth — an inversion of the magnetic body of the earth. It has already happened several times. By drilling into the earth's crust with magnetite of iron, scientists have found past cycles of North-South orientation, with evidence of as many as 20 inversions. The cycle is long — some people say 400,000 years. At the beginning of the 20th century, Steiner wrote, "Whether it takes one million years or whether it happens quickly, it is unavoidable that the North Pole will become the South."

It is clear that by ignoring the energetic "bodies " of the earth, we are enormously shortening this cycle. Today exotic fish have even been spotted off the Brittany Coast. Sudden changes in temperatures, rainfall, etc., all underline the disorganization of energetic matters on Earth; these imbalances also explain a rise of potential violence in man. One sees apparently reasonable people transformed into murderers, without understanding what has happened.

By putting biodynamics into practice, however, one can begin to address this problem: one links the Earth back to its essence through an energetic plan, and one directs human thoughts toward the forces of life in a conscious way. In this domain, quantity is of little importance. A few hectares farmed biodynamically are sufficient to reconnect much larger areas, especially if they are applied on the "acupuncture points" of the Earth, those where Cathedrals were built, or even earlier where menhirs and dolmens were placed (and where now sometimes appear, whether by design or randomly, nuclear power stations or huge television antennas).

But take care, in this return to an understanding and use of the laws of life, not to get lost, not to take flattering short-cuts. Let's take the example of Professor Bentvenis: a few years ago he demonstrated that water molecules have a "memory." This brought about an outcry from the scientific world, and a severe condemnation from the well-known magazine *Science*. What was he claiming? Only that each substance receives and emits specific frequencies, and that one could therefore produce the effects of a substance by simply holding back its frequencies and leaving the substratum. In turn, one could pass these frequencies on to water, which thereby became "informed," that is, it retained the effects of the substance in question. Water, therefore, has a memory! This would explain the healing qualities on one organ or another of springs in different parts of the Earth, each having "memorized" one or more frequencies. Bentvenis patented his discovery and has just sold it at a good price, for its commercial applications are considerable — the financiers understand that!

We now need to make use of these frequencies or microfrequencies while leaving aside the substance. We thereby leave the world of the living matter to fix on its backdrop, but it seems that we are too eager and apply this rule to any kind of substance. This recent recognition by the scientific community of knowledge which has existed forever opens the door to anything if one acts without awareness, that is to say, without consciousness of the elaborate organization of the energetic domain. One might use, for example, the energy of very toxic substances — we would then obtain water that is physically clean but charged with bad energy, therefore dangerous. In France we see many water towers covered by antennas, imposing a range of rhythms foreign to life on this water, which then passes on the negative energies to the unsuspecting people who drink it.

And one cannot help having some worries when one sees the increasingly tyrannical aspects of our global economy, which forces each person to make compromises that gnaw away at the laws of life. One of these days these new and very toxic molecules will surely be promoted as "organic" and therefore harmless. It's time that the scientists should wake up and move out of their airtight, traditional way of thinking. A brave European member of Parliament recently allowed publication of the results of her blood test, which revealed that her blood contained 50 toxic molecules from the food chain.

With deeper understanding of energetics, it will be possible to give an ecological aspect to the most dangerous substances, by putting their "emissions," or their energetic imprints, onto a neutral carrier. And it is doubtful whether their reactions with the human energetic field would even be monitored. One would then risk entering an era of considerable energetic pollution, in which the legal element would no doubt be nonexistent for a number of decades. In fact, this is something that is already happening.

This risk also applies to biodynamics. Someone has already thought of replacing the cows' horns used in two preparations with energetically imprinted porcelain. One can also imprint frequencies found at Chartres Cathedral, for example, to a newly-built concrete cellar through tiny quantities of water. The machine that reproduces these frequencies has been on sale for some years. Such examples will probably increase, and will be sold to well-meaning people. "Informed, energized" water is already on sale for thousands of euros per liter. Just a few drops are enough to "boost" the contents, say, of a bottle of wine for a number of weeks, and help it obtain brilliant good notes.

The fundamental question concerns what to keep and what to let go when one "records" the energetic emissions of a product. Is the copy absolutely identical to the original product in terms of quality? The voice of a tenor, violin music, the pleasure of an encounter, are they entirely preserved on a CD? Can Chartres Cathedral be put in a box?

In the case of a human being, the energetic emissions of this living organism include thoughts, strength of heart and will, and for lesser-evolved living matter, they include the rhythms, the inheritance of archetypal forces.

Any copy of these frequencies is in fact limited to the energetic aspect on the physical level. Although there is something missing on the recording, it does affect the physical level. It acts rather like a double, a shadow. Mankind is still very far from understanding this knowledge and is not yet ready to cope with some of these inventions.

All these examples can help explain the value of practical experience in biodynamics. "Dynamizing" with a machine, for example, might make more sense if someone is there during the process, intensely present, rather than a manual dynamization carried out by an unsatisfied employee, or by an automated machine with a timer.

This also underlines the link that a vine-grower can establish through his mind, his heart and his will with the living world that surrounds him. The vine and the Earth cannot hope for better recognition than from mankind. This link can come from different levels, for example through what have traditionally been called the "elementary beings" — as in shamanism, nowadays increasingly studied — or by rediscovering the energetic world in a much deeper fashion.

All this will be proved by tomorrow's science as soon as researchers start to look at the macrocosm and not just the infinitely small. This is the only way that man will be able to link up with plant life, for example, with its immense diversity. Each element plays a role, much like separate organs that work together in our bodies, and each plant, with its double polarity (cosmic/terrestrial) brings life forces to Earth. This is also how man can understand his role as regards animal life, with each subject acting as a sort of organ through which the Earth feels or perceives. Together they form what one could call a "body of perception" or an "astral body."

By modifying the food that animals eat by instinct, or by modifying their environment, or in a more radical way by attempting to modify them directly through genetic manipulation, one in fact diminishes the Earth's whole system of perception — just as a weed-killer destroys an immense variety of microorganisms and thus reduces the ability of the vine's roots to pick up on all the soil's subtleties. The more the diversity of animal and plant life disappears (15 million species have disappeared since 1900), the more the Earth will waste away. And man? Without proper consciousness, he will become increasingly isolated and self-absorbed. His quality life will become increasingly empty, even if by artifice he will be kept alive for longer. Man's increased sterility is a very clear sign of this deterioration.

All of this is latent in biodynamics, Waldorf's teaching and anthroposophical medicine. It is with this knowledge that one will be fully able to "bottle" the real riches of the Earth.

Conclusion

Certain winegrowers are still convinced that it is sufficient to protect their vines from disease and maybe scratch the soil a little now and then to elaborate a sincere wine. They remind me of tourists who think that choosing an expensive hotel will generate a great and beautiful trip and have missed the point of what is essential. No one can pretend to have a quality product while limiting his efforts only to the external level and concealing, consciously or unconsciously, the impact of the formative and organizing forces which descend from the macrocosm to inspire the infinite diversity of life forms. Matter does not create the forms, it harmoniously fills the contours organized by the action of intangible forces. These forces alone have the power to coordinate the elements of a much more vast organism. They are the guardians of a harmonious whole. In other times humans were born with an intuitive sense, a different perception. They saw or felt what was not directly accessible to their physical senses. The history of civilizations demonstrates this to us. We have to wait until the Greeks before artists condescend to portray physical human beauty.

Before this epoch, a conception that was only symbolic was maintained. It is obvious in the Egyptian civilization. The material aspect was not considered to be interesting. On the contrary, it was even despised. Attention was then turned toward the beyond, toward what one felt to be a formidable inspiration, like the origin of life. It was over the course of time that living oral traditions, which were still active among the Greeks, were later imprisoned in written laws whose rigidity and number only increased the more that human perceptions were stifled. Laws progressively replaced our consciences. This clearly started with Roman law. However, in the rural world, among those who had been born into the wisdom of nature, those who had been educated by the knowledge of their parents and grandparents, some of these atavistic abilities continued to survive. The sailor could find a far-away island without instruments. The wood-cutter knew when to cut the oaks to make them naturally impervious to rot. The farmer knew the propitious moment to plant the wheat in order to have healthy growth.

Many of the old peasants instinctively knew that beyond the physically perceptible world, the earth as a planet is alive. Their understanding of agriculture was always marked by this intimate and secret conviction. They were confronted with the industrial era and the almost total materialism of an educational system devoid of any artistic sense. The wars completed the extinction of this long-lost knowledge. In particular, the first World War, the causes of which are not as clear as is generally believed, killed a large proportion of the agricultural community in Europe, thus the soul of the farmer. Agriculture and wine growing based only on the physical quantitative world was imposed by commercial mentalities, for whom the profound essence of things is never taken seriously. And it was to remedy this grave error, although it had only just begun in 1924, while staying with Count Keyserling, that Rudolf Steiner gave in his *Agriculture*, the basis for a new kind of agriculture in which reasonably priced quality could be reborn. Each of us should understand that our agricultural actions, from the most important to the apparently most insignificant, are part of a permanent interaction with a vast living system. Thus each of our gestures has a meaning and generates our responsibility. The earth has carried humanity for a long time, now it is up to humanity to carry the earth. If people of the earth do not recover an understanding of their environment, their desire for authenticity will never be fully satisfied.

The "false progress" of agriculture has always taken the opposite road by separating life into pieces and bringing about the isolation of each type of vegetation while selling "assistance" to these plants in the form of chemical products. These products, from the start out of balance, wind up sooner or later on the dishes we eat from. This type of food encourages the individualism of each of our organs. And we know about the damage caused to the human organism by isolated cells that are out of control and dissociated from the rest of the cells. Their reciprocal communication no longer exists. They free themselves from the entity we consist of. If we do not take action, in a very short time the terrifying predictions of the World Health Organization concerning cancer will come true. And often the medical methods actually in use against the plague of cancer, whose origin is not understood, separate the patients even further from their living environment. Wine undoubtedly has a mission to accomplish. It is part of our daily diet, of course, but it nonetheless occupies a privileged position. A privilege that winegrowers should use as an asset to open up a new way toward the quality and authenticity of their products.

Our planet is not a young girl. She no longer has the strength to revive her mantle of vegetation, in which life expresses itself more and more weakly, without our help. Layers of calcium several hundred yards high and veins of coal which are sometimes several dozen yards in thickness testify to an intense but very far-distant proliferation of life. It is now probable that our largest forest, if it was buried by a cataclysm, would produce a vein of coal only a few inches thick. The fate of the earth depends more than ever on the human conscience. We can either act artfully or not. Humanity has the choice; we can agree to look further ahead by allowing these supersensitive forces, these creative influences, to enter our field of knowledge or we can persist in our ignorant mistreatment of our planet. This mission is illustrated by a 15th century painting, on display at the Basel museum, in which a man carries the terrestrial globe on his back. The work may have been premonitory.

But humanity refuses wisdom in spite of repeated warnings of always greater catastrophes. Nuclear explosions, the rise to power of chemical agriculture, millions of acres rendered barren and salty by herbicides and artificial fertilizers, and now genetics, engender many destructive forces which isolate the earth even further from its biosphere which should nourish it. What if our living planet revolts against this forced separation from the solar and stellar systems? Try to see electromagnetic pollution as an immense spider web which becomes more dense day by day, like an opaque veil capable of disturbing the equilibrium of the poles and our magnetic field.

A bit of humility is what humanity needs. Remember that a crack in a nuclear reactor was a probability that our leading scientists and politicians categorically rejected. And then there was Chernobyl, whose effects we are still dealing with. At the beginning of World War II, was not the Maginot Line presumed to be invulnerable? It lasted only a few weeks. France still remembers. How would the world today cope with a major earthquake? There no longer exists a city or a village capable of sustaining its autonomy, not even to the point of providing water fit to drink. There is not one well, not one river, not one spring that is not either polluted or exploited commercially. The artisans have been replaced by machines and the wise men by computers. Electronics set the rhythm of our daily lives in the banks, in the hospitals, and in our weapons systems. Imagine the consequences of a large-scale breakdown without smiling.

Biodynamic agriculture directly addresses the formative forces and is the *only* system to do so. It reinforces the life of the earth, whose vegetation could be considered as organs connected to the solar system. By concentrating the action of these forces, it obtains tangible results in spite of the small quantities used. Of course, it must be done by human hand. The cost of manual labor is higher, but is this a defect at a time when unemployment is spreading with the rapidity of gangrene? Besides, the substantial savings realized on expenses for agrichemical products should allow us to very rapidly restore a financial balance.

We know the effect of the rise to power of an agriculture based on forced assistance in which one treatment calls for another one in a vicious circle. For the last fifty years, expenses for chemical products have kept getting larger. The earth has been drugged. But it is not a question of lifting the veil on another agricultural method that is less profitable for a system habituated to living off its own imbalances. Reconversion is within the reach of wine-growers who have known how to retain a margin of freedom in their dealings with banks, who did not decide during the euphoria of the 1980s to become involved in promising loans and whose culmination now seems sometimes more dubious.

Practicing biodynamics is also accepting familiarity with another language, with other methods and with seeing during the first years a drop in production or an increased risk of disease. Reconversion does not affect only the domain, but also the individual whose motivation should go beyond simple commercial objectives. Finally, but from this one does not die, one must sometimes expect to be ridiculed.

He who chooses this conversion evolves between two major risks. The first one consists in applying biodynamics blindly and mechanically, as if it was a recipe. This static conception refuses the mark that the consciousness of each one of us must imprint on his domain. It discourages the research required to understand the location one lives in and risks creating an unhealthy dependence on a counselor, whom one can always turn into a guru. The other and opposite risk is to do no matter what and to destroy the subtle effects of the prodigious nugget of wisdom contained in Rudolf Steiner's *Agriculture*. Between the two extremes it is rather an "architectural" work that is to be accomplished, a work in which each gesture has its place.

At present there is not one single agricultural school or training center capable of transmitting the bases of an agriculture that is really alive, whether it be organic or biodynamic. What we today call "biological" is reduced to the exclusion of the use of chemical products. A good approach, but it is perhaps not sufficient. We are totally underdeveloped in this respect and we think we are being revolutionary by putting up here and there billboard signs about recuperating the byproducts of toxic treatments. Is this progress, or a false road which cleverly disguises the absence of a genuinely qualitative approach? In the sad shipwreck of present-day agriculture, the winegrower may be one of the last oases for exercising the right to quality, but this privilege will not last forever. If the winegrowers delay too long the moment of becoming aware, they will be caught in the gears of this enlarging desert of poverty. It is unavoidable, because a sickness, no matter what part of the living body it affects, if not healed can bring about the total destruction of the body. And wine growing belongs to the agricultural organism. Therefore is it not up to winegrowers to be the first to set the example for all of agriculture?

Appendix I
Rudolf Steiner

Born in 1863 at Kraljevec, in what is now Slovenia, Rudolf Steiner went on to technical and scientific studies in Vienna. In 1891 he obtained his doctorate in philosophy at Rostock. He was the youngest collaborator on the Goethe archives at Weimar where, between 1890 and 1897, he was in charge of editing the scientific writings of Goethe. As an editor, writer and lecturer, he taught at the popular University of Berlin.

He founded a modern way of approaching spiritual realities called Anthroposophy, which he described in his books and in nearly 6,000 lectures delivered all over Europe to a wide variety of audiences.

He conceived the idea for and constructed the Goetheanum at Dornach, near Basel, which was simultaneously a university, a research center, and a theater. He innovated and renewed many areas of social life including sociology, pedagogy (Waldorf Schools), healing pedagogy, medicine and pharmacy (Weleda), bio-dynamic agriculture (the Demeter label), architecture, the theater, etc.

He died at Dornach in 1925.

His complete works in German (writings and lectures) consist of 350 volumes, a large number of which have been translated into many other languages.

Appendix II
Concerning Mad Cow Disease

From a lecture given by Rudolf Steiner on Jan. 13, 1923, at Dornach. See Santé et Maladie, *by Rudolf Steiner (Editions anthroposophiques romandes, Geneva, 1983).*

"You know well that animals exist which behave completely like good vegetarians. Some animals do not eat meat. Let us take for example our cows, they do not eat meat. Neither do horses hunger for meat, they eat only vegetation. It must be understood that the animal does not content itself by ingesting food, but is continuously eliminating what is in its body. For example, you know that birds molt. They lose their feathers and must replace them with new ones. You know that stags shed their antlers. As for yourselves, you know that when you cut your nails, you notice afterwards that they grow back. But what appears in this case so visibly is happening constantly! We are constantly eliminating our skin. I have already told you about this previously. Within the space of seven or eight years, we have eliminated our entire body, which we have replaced with a new body. This is also the case among animals.

"Let's pause for a moment over a cow or a steer; well, if you take it a few years later, the flesh has completely changed. It is somewhat different with cattle than with humans: regeneration takes place more rapidly in cattle. Their flesh is thus regenerated. But what is at the origin of this flesh? That is what you should ask yourself. Pure vegetable matter is at the origin. The steer has itself produced its flesh from vegetable matter. That is the most important point that should be made. The body of an animal is thus able to transform vegetation into flesh. Well, gentlemen, you can cook a cabbage as long as you want to, but you will never obtain meat from it. You won't be able to obtain meat from it by putting in your frying pan or casserole, any more than it is possible to transform a cake you have prepared into meat. There is thus no technique to make this possible. But in summary, what cannot be done with technology can be done in the animal's body. It is simply meat that is produced in the body of the animal. But the forces necessary for this operation should already be present in the animal's body. Among all the technology at our disposal, there is not anything which transforms vegetables into meat. We don't have this. Our bodies as well as animal bodies thus contain forces capable of transforming vegetable substances, vegetable matter into flesh.

"Let us consider a plant that is still in a meadow or a field. Until now forces have acted, making green leaves grow, berries, etc. Suppose now that a cow eats this plant. A cow or steer that eats this plant transforms it into flesh. This signifies that the steer possesses within itself forces which permit it to transform the plant into flesh.

"Imagine a steer wanting to say to itself, *I'm fed up with grazing on herbs. Some other animal can do that for me, and I can then eat that animal!* See, the steer would then start to eat meat! It is nevertheless able to make meat itself! It has forces at its disposition permitting it to do this. What would happen then, if instead of vegetation, the steer started to eat meat? All the forces in it which can produce flesh would find themselves idle. Take any factory that should produce something, and suppose that you produce nothing, but that you put the whole factory to work — imagine a little the waste of energy there could be. A considerable force would thus be wasted. But, gentlemen, the force that is wasted in the body of an animal cannot be dissipated like that. The steer is overflowing with this force, which does something else in it than transforming vegetables into meat. The force remains and is certainly there, but acts on it in another way, producing in it all sorts of garbage. Instead of flesh, harmful substances are produced. If the steer suddenly becomes carnivorous, it fills up with all kinds of harmful substances, particularly uric acid and urate.

As for urate, it has its own particular habits. The particular habits of urate are an attraction for the nervous system and the brain. If the steer ate meat directly, the result would be secretion of urate in enormous quantity. The urate would go to the brain, and the steer would go crazy. If we could do an experiment of feeding a whole herd of cattle by suddenly giving them doves to eat, we would wind up with a herd of cattle gone completely mad. It is thus that this presents itself. In spite of the gentleness of the doves, the cattle would go mad."

Appendix III

The Waldorf Curriculum

Preface to the catalogue of the Waldorf Exposition for the Rudolf Steiner schools, presented in the context of the 44th session of the International Conference on Education of UNESCO in Geneva.

Pluralism in teaching and the right of parents to a free choice of school are the expression of a liberal democratic society. Such a society permits the existence of autonomous pedagogic methods and stimulates constructive competition within the educational system. The independent schools, like the state schools, fulfill a mission of public education. A liberal state should then protect and support a pluralized and public system of education. The scholastic monopoly of the State is obsolete.

In autumn 1994, the International Movement of Waldorf Schools will celebrate its 75th birthday. The interest aroused by this pedagogic model has considerably increased on the international level. Two generations after the foundation of the first Waldorf School by Rudolf Steiner, approximately two thousand institutions — in the pre-school, school and university domains, as well as in specialized education — are at work all over the world on the base of this pedagogy founded on the evolution of the child. They are counted among the best pedagogic instruments in the service of all children of all cultures and all religions for their personal and social development.

The social engagement of the Waldorf teachings in the regions at present in crisis deserves particular attention. They support the cooperative types of projects in which kindergartens and Waldorf Schools are often featured in pilot programs dedicated to pluralism in education and scholastic self-management. Throughout the world these free institutions receive the support of the Friends of the Art of Education of Rudolf Steiner.

The multiplicity of contemporary interrelationships reflects the pluralism of systems of thought and of the abilities and realizations of contemporary humanity. The development of a society depends on that of the individuals it is composed of, on the cultivation of their faculties of self-determination and social responsibility. The Waldorf pedagogy wants to be a free pedagogy, which works in consequential fashion on the individual development of human skills, based on an objective psychological evolution.

The association, Friends of the Art of Education of Rudolf Steiner, desires to bring its activities to the attention of the whole world through a traveling exposition on the international Waldorf pedagogy and its accompanying catalogue.

I wish for this enterprise to attract attention, recognition and support for the well-being of the children of this world and hope that in this way an impetus that carries with it the future can be given in the field of teaching.

Dr. Hildegard Hamm-Brücher
Minister of State of the Federal Republic of Germany
Munich, September 1994

Additional Sources of Information

Santé et maladies, Dr. V. Bott
De l'Homéopathie à la biodynamie, L. Boehrer, Nature et progrès
Le Sol, la terre et les champs, Claude Bourguignon, Sang de la terre, 1995
Jardiner avec la lune, Xavier Florin
Guide pratique de la méthode biodynamique, H. Kabish
Les Enveloppes des préparations biodynamiques, W. Konig
Les méthodes biologiques appliquées à la vinification et l'oenologie, M. l'Eglise, Editions Courrier du livre, Paris
Les Plantes, les préparations biodynamiques, W.-C. Simonis
Calendrier des semis, Maria Thun
Les Mauvaises herbes (les incinérations), Maria Thun
Pratiquer la biodynamie au jardin, Maria Thun
L'Evolution de la terre, G. Washmut
La Ferme biodynamique, F. Sattler and E. von Wistinghausen

The following titles referenced in the text are available in English:

The Plant Between Sun and Earth, Adams and Whicher
The Living Earth, Walter Cloos
The Plant, Vol. I, A Guide to Understanding Its Nature, Gerbert Grohmann
The Plant, Vol. II, Flowering Plants, Gerbert Grohmann
Agriculture of Tomorrow, Dr. E. and L. Kolisko, Kolisko Archives Publications, Bornemouth, England.
Healing Plants, W. Pelikan
Biodynamic Agriculture, Introductory Lectures, Vols. I & II, Alexander Podolinsky
Sensitive Chaos, T. Schwenk
Agriculture, Rudolph Steiner
Working with the Stars, Maria Thun

For more information on biodynamic agriculture contact:

Soin de la Terre
Vieux Serrant
49170 Savennières
France

Soin de la Terre (Care for the Earth) is a non-profit association created to help the development and putting into practice of biodynamic agriculture. It sponsors conferences and seminars on the subject, and organizes roundtable discussions and visits to farms and issues advice for those who wish to become involved with biodynamics in France.

Biodynamic Farming and Gardening Association, Inc.
Building 1002B, Thoreau Center, The Presidio
P.O. Box 29135, San Francisco, California 94129
phone (415) 561-7797, fax (415) 561-7796

The Biodynamic Farming and Gardening Association publishes and sells a large selection of books and information regarding all things biodynamic as well as a journal, *Bio-Dynamics*. The group also sponsors conferences and meetings.

Acres U.S.A.
P.O. Box 91299
Austin, Texas 78709
Phone (512) 892-4400, Fax (512) 892-4448
toll-free 1-800-355-5313, e-mail info@acresusa.com
website www.acresusa.com

Acres U.S.A. is the publisher of North America's leading journal on ecological agriculture. Appearing monthly for 30 years, *Acres U.S.A.* covers all methods of commercial-scale ecological agriculture from biodynamic to organic, sustainable to least-toxic. They also publish books on the subject as well as distribute titles on the subject from around the world. Their mail-order book catalog features many of the titles listed above and hundreds more.

Index